Social Constructions of Water Quality in South Africa

Suvania Naidoo

Social Constructions of Water Quality in South Africa

A case study of the Blesbokspruit River in the Context of Acid Mine Drainage Treatment

Springer

Suvania Naidoo
Department of Development Studies
University of South Africa
Pretoria, Gauteng, South Africa

ISBN 978-3-030-98236-2 ISBN 978-3-030-98237-9 (eBook)
https://doi.org/10.1007/978-3-030-98237-9

This Springer imprint is published by the registered company Springer Nature Switzerland AG
The registered company address is: Gewerbestrasse 11, 6330 Cham, Switzerland

This book is dedicated to the memory of my father Preggy T Naidoo. For all the moments I treasure, all the lessons you taught and all the knowledge you imparted. You are always remembered, a man of many talents.

Preface

The Blesbokspruit was declared a Ramsar site in 1986 and is one of the largest wetlands in southern Africa. The wetland acts as a purifier of mining, industrial and wastewater treatment works discharges. Its ecosystem provides shelter to hundreds of bird species. The Blesbokspruit, known as a seasonal river that dries up during winter, no longer runs dry due to mining and industry pumping wastewater into the spruit. The South African government's intention for implementing the short-term treatment (STT) of acid mine drainage (AMD) was to ensure that less risk be posed to the quality of its water systems, as opposed to the impact that untreated AMD will have on the water quality of our rivers. Further, the declining water quality of rivers has wide-ranging impacts on those dependent on that water source for their livelihoods.

This book details how the water quality of the Blesbokspruit River in Gauteng, South Africa, was socially constructed by stakeholders and key individuals in the context of AMD and its treatment. Three objectives guided this research: first, exploring the different perceptions of various stakeholders on the water quality of the Blesbokspruit, in the context of AMD and its treatment; second, identifying and explaining the contributing factors, processes and contexts that play a role in the varied social constructions of the water quality of the Blesbokspruit; and, third, exploring how the social constructions of water quality are linked to power relations in respect to AMD and its treatment. Social constructionism was used as the framing for this research to explain how water is intrinsically social.

The eastern basin acid mine drainage treatment plant was launched in February 2017. The STT of AMD involves a high-density sludge management system, and the purpose of the treatment is to discharge improved (neutralised) quality of water into the Blesbokspruit, a tributary of the Vaal River in the Gauteng Province of South Africa. When the plant was launched, stakeholders became aware of changes that were made to the STT after the public consultation process of the environmental impact assessment (EIA), as required by law. This involved a change in the sludge disposal site, for which an EIA was conducted. According to the authorities, as an emergency measure, the mine void was selected as the alternative site, without an EIA having been conducted for this site, and without knowing the potential impacts on the quality of surrounding water. This concern by some stakeholders was coupled with the fact that the neutralised water discharged into the Blesbokspruit contains high sulphate levels, which negatively impact the water quality.

Findings of the research show that stakeholders are aware that the changes in the physicality of the Blesbokspruit resulted from human interventions and varied uses of the water over the years. Such knowledge and various factors (such as the historical context of mining, current coal mining, flows and volumes of water, technology used and processes followed, information and communication, and vested interests) influence social constructions of the water quality. What counts as truth about water varied, depending on the perspective of who was talking about it, their purpose, and their individual interests. Further, how one defines water quality determines and influences what treatment processes are used to improve water quality. The book explains how these social constructions are entrenched in power relations regarding AMD and its treatment and looks at how power was used to influence decisions and to improve the water quality of the Blesbokspruit. The book explains, through the illustration of constructions, why a treatment process meant to better the water quality gained a bad reputation because of the silo effect. This case study should be used to understand what happens when government works in silos, and the impact this has on water management and water governance.

Keywords Acid mine drainage; Short-term treatment; Water quality; Environmental impact assessment; Social construction; Power

Pretoria, Gauteng, South Africa Suvania Naidoo

Acknowledgements

This book has been developed from my doctoral thesis. I would like to thank Springer for allowing me to publish this book.

It is with great pride that I acknowledge the following people that walked me through this journey. My gratitude to each one of you is boundless. Professor Dirk Kotzé, you have been my mentor since I began my research in this field, way back in 2011. I am exceptionally grateful for your support throughout this study and in my academic career. Dr Carina van Rooyen who supervised my doctoral study and Dr Naude Malan for all the support you provided.

To the participants, your remarkable stories, deep knowledge and infinite passion for what you do made all this possible: Mr Bashan Govender, Mr Andrew Barker, Mr Philip de Jager, Mr Mellerson Pillay, Dr Johan van Niekerk, Mr Charl van der Merwe, Mr Ewald Meyer, Mr Pravin Naidoo, Mr Bert Pretorius, Mr Bismark Mashau, Mr Rulani Maluleke, Mr Quinton Joshua, Mr Calvin Johansi, Mr Bertus Fourie, Ms Anna-Marie Maurizi, Ms Carin Bosman, Councillor Jill Humphreys, Councillor Wollaston Labuschagne, Dr Piet Nell, Prof Tony Turton and Mr Bennie van Zyl. I thank Mr Stan Madden, Mr Adrian Storey and Ms Sophia Tlale, for the extra time you dedicated to take me on site visits. Special thanks to Mr Stan Madden, "father of the Blesbokspruit". In my view, you are the keeper of wisdom – I thank you for stoically sharing your knowledge and for your ongoing support during this journey, and for teaching me how adventurous "birding" can be. I thank Mariette Liefferink, for rooting for me through the years. This journey had immense purpose because I observed your genuine compassion and ongoing dedication to environmental activism.

To all my special family members and friends, especially my biggest cheerleaders for the endless support: Vijay, Nilavani, Saroj, Rekha and Devi. To my colleagues and friends at the University of South Africa for your continuous support and encouragement.

Contents

About the Author

Suvania Naidoo is a senior lecturer and researcher at the University of South Africa. She graduated with a doctoral degree on 22 October 2021 from the University of Johannesburg. She began researching acid mine drainage and its socio-economic impact since 2011. Suvania presented her research on acid mine drainage at four international conferences: in Japan, Croatia, London and Milan. She first published with Springer in 2015, an article entitled "An Assessment of the Impacts of Acid Mine Drainage on Socio-economic Development in the Witwatersrand: South Africa". In 2017, she launched her first book entitled *Acid Mine Drainage in South Africa: Development Actors, Policy Implications and Broader Impacts*. The book featured in the magazine *Mining Weekly*. In 2019, the book won her the principal's prize for excellence in research at the University of South Africa, which is awarded to candidates under the age of 35 years, whose research is internationally recognised and has made a significant impact in society. She was invited to speak at Sustainability Week 2017, an internationally recognised event to present the findings of the book, among experts in the field. In 2019, she was invited to speak at the Water Show Africa and served as an advisory board member in 2020. At this event, she spoke on some of the findings of the research conducted for this book, where she co-presented a workshop with officials from the national government's Department of Water and Sanitation and the Trans-Caledon Tunnel Authority, those responsible for implementing acid mine drainage treatment in Gauteng, South Africa. She has been quoted in the South African media on the findings of this research.

Abbreviations

AMD	acid mine drainage
ARC	Agricultural Research Council
CEO	chief executive officer
CME	compliance monitoring and enforcement
CoE	City of Ekurhuleni (previously known as the Ekurhuleni Metropolitan Municipality)
COGTA	Department of Cooperative Governance and Traditional Affairs
DEA	Department of Environmental Affairs
DMR	Department of Mineral Resources
DWAF	Department of Water Affairs and Forestry
DWS	Department of Water and Sanitation
E. coli	*Escherichia coli*
ECL	environmental critical level
EIA	environmental impact assessment
EMM	Ekurhuleni Metropolitan Municipality (now known as CoE)
ERWAT	Ekurhuleni Water Care Company
GDARD	Gauteng Department of Agriculture and Rural Development
FSE	Federation for a Sustainable Environment
ha	hectare
IAPs	interested and affected parties
IMC	Inter-Ministerial Committee on Acid Mine Drainage
km	kilometres
ℓ	litre
LTT	long-term treatment
mg/ℓ	milligram per litre
Mℓ	mega litre
Mℓ/day	mega litre per day
MPRDA	Minerals and Petroleum Resources Development Act no 28 of 2002
NEMA	National Environmental Management Act no 107 of 1998
NGO	non-governmental organisation
NWA	National Water Act no 36 of 1998

PAJA	Promotion of Justice Act 3 of 2000
STT	short-term treatment
TCTA	Trans-Caledon Tunnel Authority
TSF	tailings storage facility
UK	United Kingdom
UN	United Nations
VAT	value-added tax

Chapter 1
Overview of AMD and Its Treatment in South Africa

Abstract This chapter provides an overview of the book. It presents, in brief, the concerns with acid mine drainage (AMD), how it is defined, its impact and why the treatment of AMD was so critical to avoid further deterioration of the water quality of the Blesbokspruit. It provides a brief overview of the process followed to implement the treatment of AMD in the East Rand. This chapter further provides a justification for this research and explains why social constructionism was used as the framing to reach the conclusions of the book.

Keywords Acid mine drainage · Blesbokspruit · Social construction

"Water is increasingly considered to be the lifeblood of the planet" (Biswas and Tortajada 2009, p. 1), yet water scarcity affects more than 40% of people in the world, and by 2025 about 1.8 billion people will be living in countries that have absolute water scarcity (Forouzani et al. 2012, p. 128). In a country such as South Africa, escalating water consumption due to population growth, lifestyles and climate change is increasing water shortages in water-scarce areas (Braga et al. 2014, p. 8; Hedden and Cilliers 2014, p. 4). Such concerns about water quantity are coupled with concerns about water quality. Worldwide, up to 90% of wastewater is dumped into seas, rivers, lakes and wetlands, and is left untreated, leading to a "global water quality challenge" (UN Water 2016, p. 6). Both the increased demand for water, competition over access to available water and the increased incidents of water pollution will have a visible impact on agriculture, food security, health and human well-being (Gordon et al. 2010, p. 513; Mehta 2003, p. 4; Roberts and Barton 2015, p. 13). Water is the most disputed topic on earth; ongoing debates about water exist regarding "who should own it, manage it, have access to it, profit from it, control it and regulate it" (Strang 2004, p. 1). If water is not managed efficiently, globally, future social and economic development would be jeopardised (Biswas and Tortajada 2009, p. 1). All nations require the implementation of increasingly efficient water management policies and practices in terms of both quantity and quality.

South Africa, being a dry country, is water-stressed: by the late 2000s about 98% of the available water supply had already been used (De Lange 2010, p. 62) or

S. Naidoo, *Social Constructions of Water Quality in South Africa*,
https://doi.org/10.1007/978-3-030-98237-9_1

wasted (Turton 2015, p. 15). The major causes of water contamination in South Africa include untreated sewage, industrial waste, water pollution from agricultural products such as pesticides and gold- and-coal mining activities that pose enormous threats to the quality of the water (Turton 2015, p. 26). Increased incidents of water pollution simply worsen this situation. According to Turton (2015, p. 26), the greatest failure of the South African government since 1994 has been declining water quality due to great failures in the management of municipal wastewater treatment plants. In line with this, Taylor (2016) stated that the water quality in South Africa had dropped by 8% since 2012 due to wastewater treatment plants not functioning adequately and increasing occurrences of raw sewage spills into dams and rivers that are used for drinking purposes. Furthermore, according to the South African Department of Water and Sanitation (DWS) (2018, p. 2), over 50% of South Africa's wetlands have been lost and 33% are in a poor ecological state. This is due to the poor, critical or dysfunctional condition of wastewater treatment works and water treatment works. Due to these conditions, in 2018 the DWS estimated that R33 billion more was needed each year for the next 10 years to achieve water security (DWS 2018, p. 2). For this to be achievable, a key ongoing challenge and strategic focus of the National Water and Sanitation Master Plan (DWS 2018, p. 2) is to actively address the deterioration of water quality and restore the integrity and effectiveness of water governance, delivery, management and monitoring systems. Acid mine drainage (AMD) has contributed to the deterioration of the water quality in the Blesbokspruit (McKay and Milaras 2017), a tributary of the Vaal River, in Gauteng Province, and is the specific focus of this book and interest of this research.

AMD is an age-old issue that has negatively impacted the environment in many other countries in the world, such as Australia, Canada, the United States and Germany. The issue of AMD in South Africa dates as far back as 1903 (Liefferink 2012), resulting from abandoned mine sites (Adler et al. 2007, p. 34). In South Africa it was reported in 2015 that AMD accounted for 15% of water pollution, with industrial waste, urban wastewater and agriculture accounting for the remainder (Creamer 2015). "Mariette Liefferink was one of the activists who sounded the alarm about the threat of acid mine drainage" and has continuously raised public awareness on the risks of this polluted water seeping from mining areas in Gauteng's rivers (Bega 2021, p. 13).

AMD pollutes both ground and surface water supplies as acidic water and toxic metals from decommissioned coal and gold mines filter into rivers (Hobbs et al. 2008, p. 417). A common interpretation of AMD concentrates on only "underground water decanting from mine shafts which oxidises with the sulphide mineral iron pyrite [which] then decants into the environment in the form of AMD" (Naidoo 2017, p. 20). AMD is thus formed when underground mine water that rises to the surface becomes acidic and rich in sulphate salts, occurring in areas where there are abandoned mines and underground mine water is no longer pumped to the surface (Naidoo 2017, p. 20). This is the definition used by the South African government (Coetzee et al. 2010). The book entitled *Acid Mine Drainage in South Africa: Development Actors, Policy Impacts and Broader Implications* led to the development of an extended way to define the concept (Naidoo 2017 p. 44). This book explained how the concept is defined in various ways by key stakeholders, linked to AMD and its impact on water quality, in South Africa. A broader definition, but seldom referred to as a way of defining AMD, originating from Anthony Turton, included "acid rain as a promoter for AMD originating from tailings dams. This form of AMD centres on the formation of mine residue-induced water contamination above ground, and is the link between acid rain, uranium mine dust, and radioactivity" (cited in Naidoo 2017, p. 44). The mine dust on the tailings dams includes radioactive particles such as uranium which intensify the AMD effect. The way in which AMD is defined depended on stakeholders' particular interests and institutional affiliation. This hinted at the idea of social constructionism, which is the key idea of this research and will be unpacked in this book. The book is not aimed at reaching a consensus on the definition of AMD, but rather the detrimental impact that AMD has on water quality, leading to the implementation of the short-term treatment (STT). The aim of this research was to identify selected key individuals' and stakeholders'[1] views of the impact of the STT on the water quality of the Blesbokspruit.

AMD is a threat to South Africa's water resources because the toxic metals that are dissolved in the water can cause health-related problems for people, livestock, fish and crops that are irrigated with this water (Colvin et al. 2013, p. 16). Human exposure to AMD pollutants can occur through "consumption of contaminated water, food or through dermal absorption via water or air" (Meyer et al. 2012). Long-term exposure to AMD-polluted drinking water can pose a threat to human health (Liefferink 2012; Frankson 2015; Funke et al. 2013).

Public awareness of AMD in South Africa, as seen in media reporting, has been raised since 2002, when AMD decanted for the first time in the western basin near Krugersdorp (DWAF 2009, p. 2). It reached 'crisis' proportions in 2011 due to the heavy rainfall at that time, leading to emergency plans to start pumping the mine water to keep it below the environmental critical level (ECL). The ECL indicates the level where underground AMD water starts to decant to surface water (Naidoo 2015, p. 1045). Concerns about AMD are concentrated in the Witwatersrand gold fields, Mpumalanga and KwaZulu-Natal coal fields, and the O'Kiep copper district. In Gauteng, AMD became threatening due to its impact on the three basins of the Witwatersrand gold fields (i.e. the central, western and eastern basins) situated in the economic heartland of South Africa (Coetzee et al. 2010, p. 8; Bega 2021, p. 13). In these three basins, the national government (through the DWS) has implemented the STT of AMD. Owing to the urgency this entailed neutralising the acid in the water with lime. Therefore, the water is partially treated and discharged into the nearest watercourse. The water is not yet of potable standards, which the DWS aims

[1] In this book, the concept *stakeholders* refer to individual persons, organisations, government institutions, civil society, the business sector and other organisations with an interest in the Blesbokspruit. Key individuals were interviewed for this research and are persons among these stakeholders who play a prominent role in, or have specialised knowledge on, the Blesbokspruit area regarding AMD treatment. Therefore, this book will refer to both *stakeholders* and *key individuals*.

to achieve through the implementation of long-term treatment (LTT). The STT was implemented in the western basin, after it was considered a write-off after the first occurrences of decanting in this basin (Naidoo 2017, p. 53). By the end of 2015 pumping in both the western and central basins was under control and remained below the ECL (Naidoo 2017). AMD became a concern in the eastern basin of the Witwatersrand in 2011 when the pumping of underground mine water from Grootvlei Proprietary Mines (Grootvlei Mine) stopped at the same time as the pulping plant of the Sappi Enstra paper mill was decommissioned (Ambani and Annesgarn 2015, p. 648; Ochieng et al. 2010, p. 3356). In February 2017 the eastern basin came under the spotlight through the launching of the R1[2] billion AMD treatment plant (DWS 2017a). The STT, as in the other basins, involves neutralising the acidic levels in the water before releasing this partially treated water into the Blesbokspruit (DWS 2017b; Mokonyane 2017).

The Blesbokspruit was declared a Ramsar site (discussed in Sect. 3.4) in 1986 and is one of the largest wetlands in southern Africa. It has its origins in Daveyton and Jetpark (an industrial area in Kempton Park), and runs past the towns of Springs, Nigel and Heidelberg, before joining the Suikerbosrand River. The wetland acts as a purifier of mining, industrial and wastewater treatment works discharges. Its ecosystem provides refuge for over 450 bird species (Hawley and Desmet 2020, p. 19). The Blesbokspruit used to be a seasonal river that dried up during winter, but since the 1990s it has no longer run dry due to mining and industry pumping wastewater into the spruit. The South African government's intention for implementing the STT of AMD was to ensure that less risk be posed to the quality of our water systems as opposed to the impact that untreated AMD will have on the water quality of our rivers. However, the way in which the STT was implemented and the large costs[3] involved led to much public uncertainty regarding the impact that the STT would pose to the water quality of the Blesbokspruit. While natural science studies have been done on the water quality in the Blesbokspruit (Ambani and Annesgarn 2015; Nell and Lea 2004; Nell and Schoeman 2015; Rose et al. 2007), there is limited research on the perceptions and experiences regarding the water quality of people who interact with the Blesbokspruit and the perceived impact that the implemented AMD STT poses to the water quality.

This justified the purpose of the research, which was to explore various stakeholders' perceptions of the water quality of the Blesbokspruit in the context of AMD and its STT. The STT uses a high-density sludge process, which requires a site to dispose of the sludge and the discharge of large amounts of water into the Blesbokspruit (discussed in Chap. 4). In order to dispose of sludge, in terms of South African law, an environmental impact assessment (EIA) has to be conducted to determine the impact on the environment, the water resource and the surrounding communities (RSA 1998). In 2014 an EIA was conducted for the sludge disposal

[2] R refers to the South African rand (ZAR).

[3] To clean up AMD in the short term the estimated capital expenditure is R721 million, and in the long term, R2.97 billion (Bega 2021).

site called the *Grootvlei Site 6/L/16 tailings storage facility (TSF)*, situated in the East Rand, Gauteng, and approved in 2015. Due to an unexpected change with this site, the STT had to be implemented under emergency measures. When a project is implemented as an emergency it means that the required procedures stipulated by law (those stipulated in the relevant Acts) are exempted. Therefore an alternative sludge disposal site was selected, namely, the mine void at Grootvlei Mine's Shaft No. 3 (discussed in Chap. 4). The selection of this site had to be signed off by the Minister of Water and Sanitation because of its emergency status, which exempted the project from conducting an EIA for this site. Furthermore, no EIA was conducted before the building of the AMD treatment plant and the discharging of the neutralised water into the Blesbokspruit. As a result, the voices of stakeholders along this watercourse were not taken into account. This led to much uncertainty and frustration among people who interact with the Blesbokspruit regarding the potential impact of AMD and the STT on the water quality of the Blesbokspruit.

A review of the EIA documents for Grootvlei Site 6/L/16 was conducted to identify the disputes that existed among stakeholders about the use of the site. Stakeholders (e.g. the Springs and Nigel communities, local and provincial government, researchers, activists, and the agricultural, industrial and tourism sectors) raised many concerns during the EIA public participation process. However, when stakeholders became aware of the fact that the sludge disposal site had changed to the mine void and no EIA had been conducted for the new site, they felt that their concerns had not been considered and there was no purpose to the initial EIA public participation process. This situation had already alarmed the public, and negative perceptions of the impact that the AMD STT would have on the water quality began to surface (discussed in Chap. 5).

AMD treatment requires a natural science solution, but because there are people who live and work along the Blesbokspruit and use the water for various purposes, conducting a social science analysis would be relevant to explore their perceptions and experiences of the water quality. The STT could pose negative impacts on the quality of the water which could affect people who are dependent on the water. There are a variety of stakeholders who interact with the Blesbokspruit, and, therefore, their perceptions of the water quality differ, based on their use for the water and particular interests. Hence, AMD issues in South Africa need more than a technical solution; they also require socio-political solutions (Mpofu et al. 2018, p. 78). Owing to the many activities and uses of the water of the Blesbokspruit, the AMD STT (which was meant to improve the water quality) created further concerns regarding the impact it would have on the quality of the water in an already exacerbated system.

Before the AMD treatment plant was built, concerns had already existed about water pollution incidents caused by dysfunctional wastewater treatment plants and ongoing mining operations. A range of factors then played a role in how stakeholders and key individuals viewed the water quality of the Blesbokspruit. Stakeholders raised alarming questions such as

Since no EIA had been conducted for the construction of the AMD treatment plant and the discharge of neutralised water into the Blesbokspruit or the disposal of sludge into the mine void, what impact would the STT have on an already deteriorated system?

What was the purpose of an EIA if the STT was implemented under emergency measures?

Had the government abided by the law in implementing the AMD STT or was authoritative power used?

How would new mining applications and ongoing pollution incidents be handled given the already sensitive environment?

How could the management of the Blesbokspruit be improved to account for all the activities taking place that depend on water uses?

Would the LTT also be implemented as an emergency without considering the public's concerns?

Such questions implied that there was a lack of communication with, and information for, the public regarding the process of the STT and its potential impacts on the environment.

Considering various peoples' dependence on the Blesbokspruit meant that diverse perceptions were formed on the impact of the AMD STT on the water quality, based on vested interests. Further, the way in which the STT was implemented suggested that it was linked to power relations between and among different stakeholders. A qualitative research design was used to conduct this research. Interviews were conducted with a range of individuals from various stakeholders who were prominent in the Blesbokspruit catchment and those who formed part of the project team for AMD treatment. Individuals interviewed for this research are referred to as *key individuals*, but various stakeholders' views form part of the findings of this book. The book presents the views of these stakeholders on water quality before the AMD STT was implemented (before 2016) (Digby Wells Environmental 2015) and after the STT had been implemented (from mid-2016).

Although many opinions or much speculation (gathered from the fieldwork and presented in the findings) exist about the perceptions of people who are directly affected by the water quality of Blesbokspruit, it is not based on methodical research and analysis. The research thus explored how the people who interact with the Blesbokspruit socially construct their views of the water quality, based on how they perceive the STT of AMD, and how other sources of pollution in the Blesbokspruit influence why they hold such views. Social constructionism (as a theoretical backing) is used to explain that factors influence and play a role in how stakeholders and key individuals view the water quality of the Blesbokspruit. How these social constructions of the water quality link to power relations regarding AMD and its treatment is explained. These social constructions will remain and continue to influence how one views water quality because the STT is known to have detrimental impacts for some stakeholders. Therefore the role of vested interests becomes critical because these social constructions are not uniform, and this is due to the various needs for, and use of, the water. Therefore, a strong justification for this research is to specifically address this component in the context of power relationships. At this point, what also must be strongly emphasised is that there is limited research available that draws a link between social constructions and public policymaking, which this research aimed to achieve.

The main research question that led to the findings of this book is: How is the water quality of the Blesbokspruit socially constructed by stakeholders and key individuals in the context of AMD and its treatment? Therefore, the aim is to unpack what the nature of social constructions of the water quality of the Blesbokspruit is, and what the key influences are that contribute to these constructions.

The book aimed to explore the different perceptions of various stakeholders and key individuals on the water quality of the Blesbokspruit, especially in the context of AMD and its treatment. It identifies and explains the contributing factors, processes and contexts that play a role in the varied social constructions of the water quality of the Blesbokspruit and explores how the social constructions of water quality are linked to power relations regarding AMD and its treatment.

Various studies exist on the impact of mining on water quality, caused by AMD (Mpofu et al. 2018; Moeng 2018; Rezaie and Anderson 2020). Studies have also been conducted on the government's inability to hold the mining industry accountable for the impacts AMD has on the environment (McKay and Milaras 2017; Minnaar 2020). Studies have been conducted on AMD treatment from a natural science perspective, but none from a social science perspective. This book is unique in that it unpacks the views of stakeholders on the 'perceived' impacts of the STT on the water quality of the Blesbokspruit.

The book indicates that social constructions about nature are formed from humans interacting with nature (Bakker 2003; Budds and Hinojosa 2012; Castree 2001; Chen 2013; Demeritt 2001; Linton and Budds 2014; Strang 2004; Tvedt 2015; Torregrosa et al. 2019). It implies that human-created events that occur in the environment can lead to socially produced ways that help one understand the environment in a particular way. In this specific case, the Blesbokspruit was merely a seasonal stream but due to mining activity (and wastewater treatment plants) it now has a continuous flow of water. What was known as a wetland is now a rapidly flowing river throughout the year. The water of the Blesbokspruit is co-produced by the mining industry: the Blesbokspruit is constantly flowing because of discharge of mine water, wastewater treatment works, and, most importantly for this research, this flow is maintained through another human intervention, namely, the discharge of water from the AMD treatment plant.

This case study shows how people's understanding of nature is rooted within social relations by explaining how nature is physically reconstructed over the years to serve dominant interests, such as the mining industry and that of government for the economic growth of South Africa. The book advocates that the hydrological and the social be considered together as interconnected and mutually constitutive. Instead of framing water as an '*object* of evaluation and contestation' (indicating that humans project meaning onto the world), this work contributes to shifting the established frame to foreground how the meaning of water emerges from relations with and through water (Krause and Strang 2016, p. 634). How the water quality of the Blesbokspruit is socially constructed indicates how one relates to the water, based on one's need and use for it. The uniqueness of this book is that it approaches the constructions of water quality on the basis of how the treatment of AMD is perceived, which has not yet been researched from a social science perspective.

This book aims to critically contribute to how relationships through water are construed through power relations and vice versa. The idea is that water allows one to think differently with new insights into, for example, power relations between those who manage the resource and those who use the resource. In relating ideas about water quality to power dynamics, this book aims to contribute to a fresh, critical understanding of not only society, but also of relations between society and nature, especially water and its quality. The research conducted for this book is different from other research conducted on water and power relations in South Africa because it looks at how the STT of AMD was exempted from legal requirements in order to protect the water quality of the Blesbokspruit; how power is used to influence decisions. The research has implications for policymakers and implementation of policy, such as improvement of EIA public participation processes, through strengthening information sharing and communication, and better management, such as clarified roles and responsibilities to work towards the improved cooperative government, water governance and policy to influence the social constructions of water quality.

Though this is not a book focused on policy, it can be used by government departments, such as the DWS, because it has implications for their policies and policy implementation.

The structure of the book is as follows. Chapter 2 presents how social constructionism was applied to research by discussing what social constructionism means, how water is intrinsically social and various authors' explanations of how people socially construct water quality. Chapter 3 provides the background to the Blesbokspruit Ramsar site and the eastern basin, the purpose of which is to provide context to the case study and its significance. The chapter also explains how the Blesbokspruit catchment is managed and by whom. A brief section on the research methods is provided, such as who the stakeholders and key individuals are and their specific purpose and role in the Blesbokspruit catchment. Chapter 4 provides the context for the discussion in the findings chapters of the book. An overview of AMD in South Africa, including the nature and occurrence of AMD in the eastern basin, is presented. The construction of the AMD treatment plant and the STT process and implementation are explained in detail. This chapter also presents the EIA process that was conducted for the Grootvlei Site 6/L/16 sludge disposal site to indicate how the EIA process occurred. Chapters 5–7 offer detailed discussions of the findings of the book. Chapter 5 provides an in-depth review of existing documents on the EIA process for the initial sludge disposal site; this formed a crucial part of *why* perceptions of the STT were formed. The chapter explores the various stakeholders' and key individuals' perceptions and views of AMD STT. Chapter 6 explains how the perceptions of AMD STT stem from factors that influence stakeholders' and key individuals' social constructions of water quality. What they know about the Blesbokspruit and the existing impacts on the river is *what* they use to socially construct the water quality. Chapter 7 explains *how* the social constructions of water quality are linked to power relations in respect of AMD and its treatment. Power relations are discussed in the context of legalisation, regulation and governance procedures, and issues that exist between spheres of government related to the

Blesbokspruit. The aim is to discuss how power influences the social constructions of water quality. Chapter 8 concludes the book by providing the main ideas, arguments and findings that developed from the research. Implications for future research are considered and lessons learned discussed.

References

Ambani, AE. & Annesgarn, H. 2015. A reduction in mining and industrial effluents in the Blesbokspruit Ramsar wetland, South Africa: Has the quality of the surface water in the wetland improved? *Water SA*, 41(5):648–659. Available at: https://journals.co.za/content/waters/41/5/EJC179283. Accessed on: 22 February 2017.

Adler, R., Claasen, M., Godfrey, L. & Turton, A. 2007. Water, mining and waste: An historical and economic perspective on conflict management in South Africa. The Economics of Peace and Security Journal, 2(2):33–41.

Bakker, K. 2003. The political ecology of water privatisation. *Journal Studies in Political Economy A Socialist Review*, 70(1):35–58. Available at: https://www.tandfonline.com/doi/abs/10.1080/07078552.2003.11827129?journalCode=rsor20. Accessed on: 20 May 2020.

Bega, S. 2021. Toxic mine water still a threat. *Mail & Guardian,* 28 May to 3 June.

Biswas, AK. & Tortajada, C. 2009. "Changing global water management landscape", In Biswas, AK. & Tortajada, C. (eds). *Water management in 2020 and beyond.* Springer: Berlin.

Braga, B., Chartes, C., Cosgrove, WJ., da Cunha, LV., Gleick, PH., Kabat, P., Kadi, MA., Loucks, DP., Lundqvist, J., Narain, S. & Xia, J. 2014. *Water and the future of humanity: Revisiting water security.* Lisbon: Gulbenkian Think Tank on Water and the Future of Humanity. New York, BY: Springer.

Budds, J. & Hinojosa, L. 2012. Restructuring and rescaling water governance in mining contexts: The co-production of waterscapes in Peru. *Water Alternatives*, 5(1):119–137.

Castree, N. 2001. Socializing nature: Theory, practice and politics. In *Social nature: Theory, practice and politics.* Edited by Castree, N. & Braun, B. Oxford and Malden: Basil Blackwell: 1–22.

Chen, C. 2013. Mapping water: Thinking with watery places. In *Thinking with water.* Edited by Chen, C., MacLeod, J., & Neimanis, AG. London: McGill-Queen's University Press: 274–299.

Coetzee, H., Hobbs, PJ., Burgess, JE., Thomas, A. & Keet, M. (eds.) 2010. *Mine water management in the Witwatersrand Gold Fields with special emphasis on acid mine drainage* (Report to the Inter-Ministerial Committee on Acid Mine Drainage). Pretoria: Department of Water Affairs and Forestry. Available at: http://www.dwaf.gov.za/Documents/ACIDReport.pdf. Accessed on: 24 February 2012.

Colvin, C., Nobula, S. & Haines, I. 2013. An introduction to South Africa's water source areas: The 8% land area that provides 50% of our surface water. *WWF Report.* Available at: http://awsassets.wwf.org.za/downloads/wwf_sa_watersource_area10_lo.pdf. Accessed on: 18 June 2016.

Creamer, M. 2015. Water: Decide now on Lesotho 2, acid mine drainage projects – Prof. *Mining Weekly.* Available at: http://www.miningweekly.com/article/waterdecide-now-on- lesotho-2. Accessed on: 20 January 2016.

De Lange, W. 2010. The water situation in South Africa: Some inconvenient truths. In *A CSIR Perspective on water in South Africa.* Edited by Oelofse, S. & Strydom, W. Pretoria: CSIR Natural Resources and the Environment: 62-64.

Demeritt, D. 2001. Being constructive about nature. In *Social nature: Theory, practice and politics.* Edited by Castree, N. & Braun, B. Malden, MA: Blackwell Publishers: 22–40.

Department of Water and Sanitation (DWS). 2017a. *Mine Water Management: Policy Position Draft for External Consultation and Discussion.* Gazette No. 658. Available at: https://www.greengazette.co.za/notices/national-water-act-36-1998-mine-water-management-policy-position-draft-for-external-consultation-and-discussion_20170707-GGN-40966-00658.pdf. Accessed on: 18 October 2018.

Department of Water and Sanitation (DWS). 2017b. Eastern basin acid mine drainage plant launched: Media statement. Available at: http://www.dwa.gov.za/Communications/PressReleases/2017/MS%20%20Eastern 20Basin%20Acid%20Mine%20Drainage%20Plant%20Launched.pdf. Accessed on: 22 February 2017.

Department of Water and Sanitation (DWS). 2018. *National Water and Sanitation Master Plan. Valuing Water Dignifying Sanitation*. Presentation to Provincial Consultation Workshop Preparation for the Water and Sanitation Operation Phakisa DWS: RSA.

Department of Water Affairs and Forestry (DWAF). 2009. *Development of an Integrated Water Quality Management Plan for the Vaal River system*. Water Quality Status Assessment Report. Pretoria: DWAF. Available at: http://www.dwaf.gov.za/Projects/Vaal/documents/VaalIWQMPTask4RWQOSReportFinSept2009.pdf. Accessed on: 14 April 2016.

Digby Wells Environmental. 2015. *Construction and operation of the proposed sludge disposal facility and pipelines associated with the treatment of acid mine drainage in the eastern basin of the Witwatersrand, Gauteng: Final environmental impact assessment report*. Johannesburg: Digby Wells and Associates.

Forouzani, M., Karami, E., Zibaei, M. & Zamani, GH. 2012. Agricultural water poverty index for a sustainable word. In *Farming for food and water security*. Edited by Lichthouse, E. Dordrecht: Springer: 127–155.

Frankson, L. 2015. Water management unit to deal with AMD. *Infrastructure news*. Available from http://www.infrastructurene.wswater-management-unit-to-deal-with-amd/. Accessed on: 14 August 2020.

Funke, N., Nienaber, S. & Gioia, C. 2013. An interest group at work: Environmental activism and the case of acid mine drainage on Johannesburg's West Rand. In *Public opinion and interest group politics: South Africa's missing links*? Edited by H. A. Thuynsma. Pretoria: African Institute of South Africa: 1–26.

Gordon, LJ, Finlayson, CM. & Falkenmark, M. 2010. Managing water in agriculture for food production and other ecosystem services. *Agricultural Water Management*, (97)4:512–519. Available at: http://www.sciencedirect.com/science/article/pii/S0378377409000924. Accessed on: 3 March 2015.

Hawley, G. & Desmet, P. 2020. *Review of the City of Ekurhuleni Bioregional Plan 2020*. Draft for public comment, July 2020.

Hedden, S. & Cilliers, J. 2014. Parched prospects: The emerging water crisis in South Africa. African Futures Papers 11, September 2014. Pretoria: Institute for Security Studies. Available at: http://www.issafrica.org/uploads/AF11_15Sep2014.pdf. Accessed on: 3 March 2015

Hobbs, P., Oelofse, SHH. & Rascher, J. 2008. Management of environmental impacts from coal mining in the upper Olifants River catchment as a function of age and scale. *Water Resources Development*, 24(3):417–431. Available at: http://www.orangesenqurak.com/UserFiles/File/OtherV2/Management%20of%20Environmental%20Impacts%20from%20Coal%20Mining%20Hobbs%20et%20al.%202010.pdf. Accessed on: 14 April 2012.

Liefferink, M. 2012. Environmental risks and hazards pertaining to AMD and radioactivity within the Witwatersrand gold fields. PowerPoint Presentation. Sandton: Federation for a Sustainable Environment.

Linton, J. & Budds, J. 2014. The hydrosocial cycle: Defining and mobilising a relational-dialectical approach to water. *Geoforum*, 57:170–180.

Krause, F. & Strang, V. 2016. Thinking relationships through water. *Society & Natural Resources*, 29(6):633–638.

McKay, TJM. & Milaras, M. 2017. Public lies, private and the forced closure of Grootvlei Mine, South Africa. *The Journal for Transdisciplinary Research in Southern Africa*, 12(1). Available at: https://td-sa.net/index.php/td/article/view/347/399. Accessed on: 12 June 2019.

Mehta, L. 2003. Contexts and constructions of scarcity. *Economic and Political Weekly,* 38(48): 1–12. Available at: https://www.researchgate.net/profile/Lyla_Mehta/publication/251901079_

Contexts_and_constructions_of_water_scarcity/links/560e4c8908ae96742011f88.pdf. Accessed on: 8 March 2017.

Meyer, F., van der Burgh, G. & Coertsen, L. 2012. *The impact of coal mining on agriculture: A pilot study*. Pretoria: Bureau or Food and Agricultural Policy. Available at: http://www.maizetrust.co.za/upload/WEBSITE/ResearchMarket&Production/2016/20160211BFAP%20Impact%20of%20Coal%20Mining%20on%20Agriculture%20in%20Delmas%20(1).pdf. Accessed on: 25 January 2017.

Minnaar, A. 2020. Water pollution and contamination from gold mines: Acid mine drainage in Gauteng Province, South Africa. In *Water, governance, and crime issues*. Edited by Eman, K., Mesko, G., Segato, L. & Migliorini, M. Cham, Switzerland: Springer International: 193–219.

Moeng, K. 2018. Community perceptions on the health risks of acid mine drainage: The environmental justice struggles of communities near mining fields. *Environment, Development and Sustainability*, 21:2619–2640. Available at: https://0-doi-org.oasis.unisa.ac.za/10.1007/s10668-018-0149-4. Accessed on. 1 June 2020.

Mokonyane, N. 2017. Statement by the Minister of Water and Sanitation at the launch of the Eastern Basin long-term treatment plant [Personal attendance], 17 February.

Mpofu, C., Morodi, TJ. & Hattingh, JP. 2018. Governance and socio-political issues in management of acid mine drainage in South Africa. *Water Policy*, 20:77–89. Available at: https://iwaponline.com/wp/article/20/1/77/38136/Governance-and-socio-political-issues-in. Accessed on: 1 June 2020.

Naidoo, S. 2015. An assessment of the impacts of acid mine drainage on socio-economic development in the Witwatersrand: South Africa. *Environment, Development and Sustainability*, 17(5):1045–1063.

Naidoo, S. 2017. *Acid mine drainage in South Africa: Development actors, policy impacts and broader implications*. Switzerland: Springer Nature.

Nell, JP. & Lea, I. 2004. *The effect of the Blesbokspruit wetland system on the Purification of gold mine effluent water for use by irrigated agriculture*. Centurion: ARC, Institute for Soil, Climate and Water. Available at: www.sancid.org.za/SANCID2004.htm. Accessed on: 2 November 2016.

Nell, JP. & Schoeman, JL. 2015. *The impact of water quality on irrigated agriculture along the Blesbokspruit between Herbert Bickley water care works and Heidelberg* (Report Number GW/A/96/6). Pretoria: Agricultural Research Council. Available at: http://www.researchgate.net/publication/279537012. Accessed on: 15 November 2016.

Ochieng, GM., Seanego, ES. & Nkwonta, OI. 2010. The impacts of mining on water resources in South Africa: A review. *Scientific Research and Essays*, 5(22):3351–3357. Available at: http://www.academicjournals.org/article/article1380621809_Ochieng%20et%20al.pdfAccessed on: 2 February 2017.

Rezaie, B. & Anderson, A. 2020. Sustainable resolutions for environmental threat of acid mine drainage. *Science of the Total Environment,* 717. Available at: https://www.sciencedirect.com/science/article/abs/pii/S004896972030721X. Accessed on: 10 October 2020.

Roberts, E. & Barton, B. 2015. *Feeding ourselves thirsty: How the food sector is managing global water risks. A benchmarking report for investors*. Boston: Ceres. Available at: https://www.ceres.org/issues/water/agriculture/water-risks-food-sector/food-water-risks. Accessed on: 5 January 2017.

Rose, PD., Moffett, M., Pulles. W., Nell, JP., Louw, D., Melville, A., Leucona, A., Kumalo, S. & de Wet, C. 2007. *Integrated beneficiation of mine wastewaters: Mine drainage wastewater treatment and planning for the social and labour component in sustainable mine operation and closure. Volume 2* (WRC REPORT No: TT409/09). Pretoria: Water Research Commission.

Republic of South Africa (RSA). 1998. *National Environmental Management Act no 107 of 1998*. Pretoria: Government Printer. Available at: https://www.gov.za/documents/national-environmental-management-act. Accessed on: 14 April 2013.

Strang, V. 2004. *The meaning of water*. London: Berg.

Taylor, T. 2016. SA tap water not being cleaned properly: Experts. *ENCA,* 24 May. Available at: http://www.enca.com/south-africa/sa-tap-water-not-being-cleaned-properly-experts. Accessed on: 29 March 2016.

Torregrosa, ML., de las Mercedes, D., Acosta, A. & Kloster, K. 2019. Water and society: The multidimensionality of water quality. In *Water quality in the Americas: Risks and opportunities*. Edited by The Inter-American Network of Academies of Sciences. Mexico: INAS: 20–26.

Turton, A. 2015. Sitting on the horns of a dilemma: Water as a strategic resource in South Africa. *@Liberty*, 6(22). November 2015. Available at: http://irr.org.za/reports-and-publications/atLiberty/files/liberty-2013-sitting-on-the-horns-of-a-dilemma-2013-water-as-a-strategic-resource-in-south-africa. Accessed on: 18 March 2016.

Tvedt, T. 2015. *Water and society: Changing perceptions of societal and historical development.* London, New York: IB Tauris.

UN Water. 2016. *Towards a worldwide assessment of freshwater quality* (UN-Water Analytical Brief). Available at: http://www.unwater.org/fileadmin/user_upload/unwater_new/docs/UN_Water_Analytical_Brief_20161111_02_web_pages.pdf. Accessed on: 10 April 2017.

Chapter 2
Understanding Social Constructionism of Water Quality

Abstract This chapter presents social constructionism as the theoretical framing used in this book. Research situated within the social sciences has to be driven by a theoretical framework that enables the researcher to concentrate on a social analysis of the topic. The subject matter of this book is dominated by the natural sciences and creates a stark contrast with the social sciences, which focuses on the human element. Perceptions, attitudes or constructions are core elements of the human condition. Coexistence between the human and natural sciences, between human beings and nature and between objective conditions and perceptions guided the research towards social constructionism as the foundation of this research conducted for this book. Social constructionism in the context of water is directly related to the hydrosocial cycle. The hydrosocial cycle is therefore an important concept underlying most of the discussions on social constructionism in this book. In this chapter, social constructionism is explained, followed by a discussion of the social construction of nature and the hydrosocial cycle and social constructionism of water and water quality through the use of examples that further illustrate how social constructions embed power relations.

Keywords Social constructions · Perceptions · Hydrosocial cycle · Water quality

2.1 Introduction

This chapter presents social constructionism as the theoretical framing used in the book. Research situated within the social sciences has to be driven by a theoretical framework that enables the researcher to concentrate on a social analysis of the topic. The subject matter of this book is dominated by the natural sciences and creates a stark contrast with the social sciences, which focuses on the human element. Perceptions, attitudes or constructions are the core elements of the human condition. Coexistence between the human and natural sciences, between human beings and nature, and between objective conditions and perceptions guided the research

S. Naidoo, *Social Constructions of Water Quality in South Africa*,
https://doi.org/10.1007/978-3-030-98237-9_2

towards social constructionism as the foundation of this research conducted for this book.

There are many uses for the water of the Blesbokspruit, such as for agricultural, mining, tourism/recreational and domestic purposes. Thus, there are various interests, with the water influencing how the users understand and construct water and water quality in many ways. The impact of untreated acid mine drainage (AMD) has already threatened this water source. However, because the acidic mine water is only partially treated and discharged into the Blesbokspruit, perceptions of the possible threats that the short-term treatment (STT) poses to the water quality have developed. Although the exact source of pollution that the AMD STT can cause is unknown to water users, certain issues and possibilities are emphasised, and others minimised, creating perceptions regarding the power dynamics about how the Blesbokspruit is managed (discussed in Chap. 7).

It is important to note that based on the variety of data collection methods the findings of the book represent the view of key individuals among stakeholders who were interviewed for this book and stakeholders in general linked to the Blesbokspruit. Therefore, it is important to mention that there is an interaction between these key individuals and their stakeholder grouping. This means that even though key individuals were interviewed, it is not entirely an individualised process. It is possible that these individuals' thinking about the Blesbokspruit and the treatment of AMD are linked to, and influenced by, their stakeholder grouping. Throughout the book, terms such as *perceptions*, *views* and *social constructions* are used. *Perceptions* refer to how various stakeholders view the water quality of the Blesbokspruit, whereas *social constructionism* draws on a theoretical framework to analyse how and why specific perceptions or views are produced, and how they are related to social constructionism. This frame made it possible to explore the perceptions of the STT of AMD of those who interact with the Blesbokspruit (discussed in Chap. 5) and how these perceptions influence their social constructions of the water quality of the Blesbokspruit (discussed in Chap. 6).

Social constructionism in the context of water is directly related to the hydrosocial cycle. This cycle relates to how people frame water as social and how they attribute meaning to water. The cycle is therefore a process in which social elements are essential to how we understand and interact with water. The hydrosocial cycle looks at the socio-natural processes by which water and society make and remake each other, reflecting the social nature of water. Further, this cycle explains how human understanding of water is socially produced and how water is 'known' by them. The hydrosocial cycle is therefore an important concept underlying most of the discussions on social constructionism in this book. The structure of the chapter is as follows: social constructionism is explained, followed by a discussion of the social construction of nature, and the hydrosocial cycle and social constructionism of water and water quality through the use of examples that further illustrate how social constructions embed power relations.

2.2 What Is Social Constructionism?

Wagemans (1986, p. 262) stated that the answer to the question of what is perceived tends to have an important influence on the answer to the question of how one perceives. These two questions, he claims, are usually considered as central problems for any theory of perception. In this book, *perceptions* refer to the thought processes of a stakeholder, which implies that it involves a cognitive process. *Social constructionism* refers to perceptions and views based on social settings (discussed below in more detail), and *individual constructs* have a social focus rather than an individual focus (Andrews 2012). This means that the individual's thought processes are based on a social setting. This book will refer to how individuals mentally construct their world of experiences (the concept 'perceptions' or 'views' is used to indicate this), as well as how their individual experiences can be socially produced. Each person perceives the world differently and actively creates his or her own meanings from events. Each individual develops a system of dimensions of meanings and constructs, although it is possible for him or her to show respect for others' constructions. As Ibarra and Andrews (1993, p. 278) indicated, "Perceptions are shaped by the opinions of salient or relevant others," and social constructionists stated that knowledge of the world is constructed between people (Burr 2015, p. 2). In doing this, one has the ability to change one's own constructions of the world and create new possibilities for one's own actions (Burr 2003, p. 20).

According to Burr (2015), there is no single description of social constructionism on which everyone would agree. Social constructionism "insists that we take a critical stance toward our taken-for-granted ways of understanding the world and ourselves" (Burr 2015, p. 2). Further, it enables one to question the idea that one's views of the world cannot yield its nature without difficulty and to challenge the view that conventional knowledge is based on unbiased observations of the world. Social constructionism argues that the ways in which one commonly understands the world, the categories and concepts one uses, are historically and culturally specific (Burr 2015, p. 2). One can agree that concepts are constructed rather than discovered yet continue to believe that they correspond to something real in the world (Andrews 2012). For example, one may view a table as being crafted from excellent wood; another focuses on its common design and still another views it as too small (Burr 2003, p. 3). All three views are based on different perspectives, namely, their quality, style and practicality.

Social constructionism implies that there is no right or wrong, or that knowledge comes from experience. One must not assume that what one observes actually exists; it implies that one should be suspicious of what one assumes the world to be, and the categories in which humans understand the world do not necessarily refer to real aspects. For example, specific categories determine what makes a person a man or a woman, such as gender, masculinity or femininity. By extension, what specific factors influence our constructions of water and water quality should be identified. Therefore, the way one understands something can be historically or culturally relative. It follows that products of a specific culture are dependent upon the particular social and economic arrangements that occurred in that culture during that time. Thus, social constructs strongly consider that one's knowledge is sustained by social processes. People construct it among one another, through daily interactions, and that is how different 'versions' of knowledge are fabricated (Burr 2003, p. 4). Thus, what is learnt from historical events or culture determines what one views as truth, and this determines many people's acceptance of how they understand the world. It is a series of social processes and interactions where people are constantly engaged with one another.

The use of social constructionism has become popular in the social sciences since the 1990s and 2000s, and scholars such as Demeritt (2002), drawing on Hacking (1999), broadly categorised such uses. Its first use is social construction as refutation, meaning that constructions are used to falsify particular claims about the world. The reason for this is to criticise taken-for-granted beliefs about the essential nature of things. Experiences or developments are not natural but socially constructed (Demeritt 2002, p. 769). The aim is to point out that once a person realises that those beliefs were socially constructed, it is within his or her power to change those beliefs. Many things are said to be socially constructed, and such things can be products of historical events, social forces and ideology (Hacking 1999, p. 2).

The second use of social constructionism is social construction as a philosophical critique. This is concerned with "situating human knowledge socially with advancing an understanding of reality or specific entities as socially produced, rather than as simply given with fixed ontological properties" (Demeritt 2002, p. 771). This form of social constructionism (as a philosophical critique of the Enlightenment's assumptions) is an epistemological tool to explain reality (Andrews 2012; Heft 1981, p. 229). Bruffee (1986, p. 774) argues that a "social constructionist position in any discipline assumes that entities normally called reality, knowledge, thought, facts, texts, selves, and so on are constructs generated by communities of like-minded peers". The implication is that constructionists view knowledge and truth as being produced socially, rather than being learnt by the mind. Similarly, Gergen and Gergen (2004) phrased *constructions* as inviting "openness to many ways of naming and valuing".

Demeritt (2002, p. 770) identified four strands of social construction as a philosophical critique, namely, phenomenological constructionism, sociology of scientific knowledge, discursive constructionism and actor network theory. All four strands "use the construction metaphor to question the culture/nature, subject/object and representation/reality dualisms that provide the conventional philosophical foundation for distinguishing true conceptions of nature from false ones" (Demeritt 2002, p. 767). Many of the discussions in this book are influenced by the phenomenological constructionism strand: the construction of social and environmental problems to explore how nature, and specifically water quality, is socially constructed. Phenomenological constructionism is sceptical about the "truth of concepts of nature and about the existence of natural phenomena, but realistic about society and social actors" (Demeritt 2002, p. 770). This strand of constructionism argues that people construct and redefine their realities through continual social interactions. This means that human knowledge is situated socially, or is socially produced, and this happens through an understanding of reality or specific entities rather than assuming that it simply exists that way. This approach understands "social and environmental problems as products of particular constructions of social reality, rather than of actual physical conditions" (Demeritt 2002, p. 771). The distinction is significant because it hints at scepticism about the existence of problems under investigation and the truth of any claims about them by the research subjects. Phenomenological constructionists prefer not to make judgements on this; "phenomenological construction takes no stand on the question of whether what is cognitively constru(ct)ed is materially constructed". They describe the world, rather than trying to judge or change it. For these constructionists, the process of construction is cognitive. This is the approach followed to understand "social and environmental problems as products of particular constructions of social reality, rather than of actual physical conditions". According to Demeritt (2002, p. 776), critics have disapproved of social construction, reducing it to "part of a generalised flight from reason and reality". This book analyses how individuals construct their own versions of the reality of how the STT of AMD was implemented in the Blesbokspruit.

2.3 Social Constructionism of Nature

Constructions are not only formed between human beings in social settings, but also between human beings and nature. This means that human-created events that occur in the environment lead to socially produced ways of experiencing and understanding the environment and vice versa (explained in detail below). Understanding the social construction of nature can assist one in acknowledging the power of humans to shape nature, both through their concepts and through material practices that lead to, and follow from, those ways of constructing nature. The social construction of nature can refer to either the construction of concepts of nature or to "the process of constructing nature in the physical and material sense" (Demeritt 2002, p. 767).

In understanding the social construction of nature, various authors (Demeritt 1999; Harrison and Burgess 1994; Whatmore and Boucher 1993) remind us that while nature cannot be viewed outside of its social relations, it also cannot be reduced to only that. Various authors (Castree and Braun 1998; Castree 2001; Demeritt 2001) have indicated that the crucial role of power relations in socially constructing environmental realities is to alleviate environmental problems and sustain environmental resources. Demeritt (2002, p. 768) suggests that arguments about constructionism do not acknowledge the physical reality of nature, which requires deep respect; "if not for its own sake, then because it will impact on us materially in ways, we will never be able to understand", or improve, if we view it to mirror social interests.

Consequently, Castree (2001, p. 3) refers to three broad approaches to the society–nature problem, namely, (1) people and the environment, and the (2) ecocentric and (3) social approach. The people and the environment approach refers to refocusing on the big questions of the present era concerning the problems and possibilities resulting from the human alteration of natural resources, environments and organisms. The knowledge produced from this approach tends to be technocratic, meaning that it seldom discusses the necessary socio-economic processes that transform nature in its present form. It tends to include knowledge leading to policy focused on improving environmental problems without ever addressing the deeper causes responsible for those problems in the first place. This interpretation leans towards knowledge by, and for, those with a vested interest in not having to pay the price for eradicating environmental problems altogether. The ecocentric approach centres on a nature-first approach, achieved through the breaking down of existing systems of production and consumption. This ecocentric approach is contrasted with the social approach. Both approaches believe that technocratic knowledge is unable to treat nature as anything other than a resource to be used, destroyed or regulated for mainly human benefit (Castree 2001, p. 4). The social approach views nature as unavoidably social. The book departs from this understanding of human–nature interactions.

Nature is seen as physically reconstructed by different societies to serve a specific and typically dominant social interest, implying that the social and natural link in ways that make their separation inescapable. In the case of the Blesbokspruit, for example, the physicality of the river has changed, from seasonal flow to a constant flow. The constant flow is due to human interventions such as water discharges from mining activities, water coming from wastewater treatment plants and the discharge of water as a result of the STT. The social and ecocentric approaches are similar in that they regard people and the environment as a hindrance to adequately understanding and altering society–nature relations. Further, the technocratic thinking assumes that it is possible to know nature as it really is, that when people physically interact with nature, they are dealing with things that are non-social. The taken-for-granted conceptions of nature among scholars around the globe suggest that policymakers, businesses and ordinary people are seen as part of the problem and not the solution, if societies are to build survivable futures (Castree 2001, p. 4).

For this reason, Castree (2001, pp. 6–10) suggests that nature is a complex concept and that it can be defined in four ways: (1) external nature, where nature is external to, and different from, society; (2) intrinsic nature, where nature is seen as fixed and unchanging; (3) universal nature, where nature is frequently seen to possess common elements; and (4) social nature. Defining nature as social has influenced this research and the findings of the book. In social nature two things are evident: first, nature has never been simply natural, but intrinsically social in different ways, at different levels and with a range of serious consequences. Second, the habit of talking of nature in itself as a domain that is by definition non-social and unchanging can lead not only to confusion but also to the continuation of power and inequality in the wider world. The social approach therefore advocates that nature has a complex social element, meaning that the facts of nature do not speak for themselves. In reality, what counts as the truth about nature varies depending on the perspective of who is talking about it, their purpose and what the individual or group interests are (Castree 2001, p. 9).

Nature is social in three related ways (Castree 2001, pp. 10–19): (1) knowing nature, (2) engaging nature, and (3) remaking nature. The first way in which nature is social is knowing nature, which implies that the claim that knowledge of nature is regularly adjusted with the biases of the knower(s). Castree suggests that there is "no singular, objective knowledge of nature, only the particular socially constituted knowledges, in the plural". This can be applied to how nature is understood by people, governments and businesses. As a result, arguments are made that are designed to draw attention away from the real problem, such as the maldistribution of resources. There are beliefs of nature that aim to protect the reality of nature and serve explicit social interests. In this way, it is not possible to change one's thinking to find out the truth versus one's beliefs in relation to nature. This is because different individuals use different discourses to make sense of the same nature. These discourses do not reveal or hide the truths of nature but, instead, create their own truths. Whose discourse is accepted as being truthful is a question of social struggle and power politics.

Chen (2013) suggests that what is the truth is part of an ongoing conversation that humans have with other living entities. Thus, when one speaker refuses to listen to the communication of another (such as concerns regarding the increasing loss of fish in the river), the rejected party is referred to as not being knowledgeable enough. "When the perceived order of social norms is achieved through a selective editing of an otherwise messy world of infinite and therefore dangerous potential, the act of denying the intelligibility of certain phenomena helps to maintain the status quo – it is from disorder that new social and conceptual patterns emerge" (Chen 2013, p. 278). Communication from speechless non-citizens such as animals, plants, water and rock is often rejected as "noisy babble", except when represented by human "experts" (Chen 2013, p. 278). This explains Castree's 'knowing nature' claim about whose claim is accepted and is reduced to social struggles and power politics. Power politics stem from various elements of power used by stakeholders (discussed in Chap. 7).

The second way in which nature is social is 'engaging nature'; this entails not disagreeing with the material reality of what is referred to as natural, such as trees, rivers and animals. It is a claim that the physical prospects and limitations of nature represent societies that can only be defined relative to specific sets of economic, cultural and technical relations and capacities. Nature will have different physical qualities (dependent on social practices, thus not fixed) and consequences, depending on how that society uses it. For example, natural hazards are defined according to the vulnerability of the different groups in society.

Thirdly, nature is social through 'remaking nature', which proposes that societies physically reconstruct nature, both intentionally and unintentionally. Western societies are now physically reconstituting nature in ways that increasingly escape their control (Castree 2001, p. 14). For example, in relation to this book, the possibility of acidic water from abandoned mines seeping into the main watercourse is a manufactured risk that is an unintentional and uncontrollable consequence of mining remaking the natural environment over many years.

Castree's (2001, p. 16) intention was to highlight the fact that nature is nothing more than a social construction, "where nature it seems, is depicted as mere putty in the hands of modern science and technology and business". This idea of nature not only amplifies the power of societies, but also questions the society–nature distinction. The key idea for the book is to show how nature (and therefore also water) is social; how the physicality of the water has changed through human interventions, yet also impacts on how people construct the water quality of the Blesbokspruit.

2.4 The Hydrosocial Cycle and Social Constructionism of Water and of Water Quality

This section explains how water is socially constructed by drawing on several authors and relevant examples. More importantly, it explains how perceptions of water lead to social constructions of water quality and are a result of vested interests, linked to, and embedded in, power relations. Several themes emerged on how people construct water quality, which is used in the findings of this book.

"Water is a total social fact"; it plays a significant life-sustaining role in ecological functioning, food production, economic activities, and health and recreation, and its importance as spiritual value makes it a resource that is significant for both nature and society (Molle et al. 2007, p. 1). A key interest in water is those interpretations that society has made of water, which sequentially shape how it comes to be understood (Krause and Strang 2016, p. 634). What one understands about the communications of different water systems is inevitably limited by one's use of languages and concepts, the tools used to create knowledge (Chen 2013, p. 275). There are cases signifying the need to explore in more detail the varied characteristics and social implications of water, also because of the natural characteristics that have allowed humans to signify the meaning of water in ways that no other element can match (Tvedt 2015, p. 7). To capture the idea that nature and society are not two distinct phenomena but rooted in each other, emerging theories propose considering water and people through their mutual interactions, suggesting that they exist and indeed 'are' because of, and through, each other.

The social construction of water – and the entangled nature of relations between humans and water – has led some to talk of the hydrosocial cycle (Boelens 2014; Linton 2010; Schmidt 2014; Swyngedouw 2009). Linton and Budds (2014, p. 175) define the *hydrosocial cycle* as "a socio-natural process by which water and society make and remake each other over space and time"; in other words, social processes shape water (and its quality), and water (and its quality) shapes social relations (Bakker 2003; Linton 2010). Krause and Strang (2016, p. 633) understand the water and society relationship as "rather than treating water as an *object* of social and cultural production – something produced through social relationships and instilled with meaning through cultural schemes – we consider water as a generative and agentive *co-constituent* of relationships and meanings in society". Likewise, hydrosocial understandings mean that water and society are internally related and co-created, and that particular kinds of social relations produce different constructs or perceptions of water and vice versa (Linton and Budds 2014, p. 175). "Water inspires novel ways of thinking about key aspects of social relations such as power, community and knowledge" (Krause and Strang 2016, p. 633). However, with different narratives, values, stakeholders and power bases for different water uses, it is difficult to deal with the resulting competition between different sectors for both quantity and quality of water (Molle et al. 2007, p. 2).

The hydrosocial cycle contemplates how water is manipulated, used and concentrated by stakeholders; and how the struggles for water access and control, and exclusion are expressed in institutions through influences such as legislation, institutions, practices and symbolic meanings. A relevant example for this book is mining, illustrated by Budds and Hinojosa (2012). To approach the relationship between mining and water, one should begin with the notion that water is not simply a material substance subject to human manipulation, "but a 'hybrid nature' in which water's materiality and its social relations constitute and express each other" (Budds and Hinojosa 2012, p. 120). Viewing water as being co-produced allows one to think about the ways in which social processes shape water and vice versa. Looking, firstly, at the flows, forms, practices and discourses that describe water in mining contexts, for instance, will reveal the material and social processes through which examples of water are formed. This entails consideration of a variety of aspects, such as "physical flows, patterns of access, technologies, institutions, practices, legislative reforms, governance frameworks, and discourses around water, which are mediated by social and political processes and collectively constitute the waterscape of a given context" (Budds and Hinojosa 2012, p. 120). This example is of relevance to the findings because in the Blesbokspruit the physicality of the river has changed from a seasonal flow to a constant flow. The flow is created by human interventions such as mining and the discharges of water from the AMD STT.

Torregrosa et al. (2019, p. 24) suggest that the flow of water is both a social and physical process. This circulation allows an understanding of how flows of water, capital and power are materially linked. Water flows entail the complete hydrosocial cycle, not only the relationships activated for its access, but also those involved in the uses, quality and ways in which its final disposal takes place (Torregrosa et al. 2019, p. 24). Further, there is a need to address who is entitled to what quality and volumes of water, and who should control, manage and decide how the hydrosocial cycle will be (Swyngedouw 2009, p. 57). According to Zwarteveen (2015, p. 9), water scholars use terms such as 'waterscapes', 'hydrosocial networks' or 'hydrosocial cycles' to show the growing pressures on water resources. They also suggest that the resulting scarcities are not natural processes, but the outcome of specific histories and practices of water resource exploitation or development.

Ohlsson and Turton (2000, pp. 1–2) explain the challenges of water scarcity, such as the commonly known perception that the "rivers are running dry"; meaning there is not enough water, or the demand and supply concerns lead to a widespread perception of scarcity. Societies use different ways to adapt, depending on the stage of scarcity and in line with the perception of what is necessary to adapt to the challenge of scarcity at the given time. Water is a public good, but it leads to a "public worry good" because of the increasing number of people it has to be divided between. The scarcity becomes a result of social misuse of the resource and widespread social disruption, indicating that "the story of meeting the challenges of water scarcity is a social story". Some believe that shared water resources tend to lead to cooperation rather than conflict between people and countries. The way one tells a story allows one to conceptualise the way one understands the challenge. Thus, "it is important to identify what the stories we tell, teach us to do and reflect upon whether it is really the best story to tell, in order to help us understand what the most urgent challenges are, and how to respond to them" (Ohlsson and Turton 2000, p. 2). Current water management practices influence how one tells a story about water and is the starting point of how perceptions develop about the extent of the scarcity. These perceptions are fuelled by the kind of adaptive measures taken, and what social tools are explored to address the challenges and the conflicts created through new ways of using water resources socially.

These adaptive measures form a 'spiral movement' that rotates between a perceived scarcity of water as a natural resource and a perceived scarcity of water as a social resource. What is required to overcome the original scarcity entails "progressing towards increased amounts of social resources applied to overcome the natural resource scarcity" (Ohlsson and Turton 2000, p. 2). This creates an understanding of the interaction between the sphere of natural resource management and the social challenges faced. In order to address the scarcity of the resource from a natural science perspective, the social issues that arise need to be addressed to overcome the actual scarcity.

Ohlsson and Turton (2003, p. 1) propose that the link between water scarcity and social instability is mostly a result of population growth, and when water is scarce, it is more valuable (Kaika 2003, p. 948). The unpredictability of drought is normally viewed as a natural phenomenon, and when it is understood through self-explanation, then it is viewed as a socially constructed phenomenon; such as water scarcity.

A number of case studies on the social construction of water have been published. For example, Mehta (2003, p. 1) shows how 'water scarcity' is constructed differently by various social and political actors, often to reach particular political ends, that is, power in terms of access, use and control over this resource. Aguilera-Klink et al. (2000) similarly unpacked the shift in Tenerife, Canary Islands, from ideas of water abundance in the fifteenth century to water scarcity in the twenty-first century. Such a shift led to a change in understanding water from a public good to a private commodity, benefiting landowners who have access to groundwater. Therefore, "differing explanations should not be sought for what is taken to be scientific truth and a scientific falsehood" (Aguilera-Klink et al. 2000). All knowledge and knowledge claims are to be treated as being socially constructed, thus, "explanations for the genesis, acceptance and rejection of knowledge-claims are sought in the domain of the social world rather than in the natural world" (Pinch and Bijker 1987, p. 401).

The following are a number of authors who illustrate this point: Mehta (2011) looked at the social construction of water scarcity in India; Kaika (2003) at water scarcity in Greece; Bakker (2003) at water privatisation in several countries;

Swyngedouw (1999) at hydraulic engineering in Spain; Oktem (2015) and Norman and Bakker (2009) at water management (especially integrated water resource management) in Turkey; Lien (2006) focused on bottled spring water in Norway; Pickering (2014) on water abundance in Dominica; and Palomino-Schalscha et al. (2016) on dam water in Chile. These examples illustrate how environmental problems are seen as products of particular constructions of social reality rather than physical conditions, as in the case of the Blesbokspruit.

One of nature's realities is water and its perceived quality. While there are numerous studies on water quality, they typically explore people's perceptions of water quality and water reclaiming (see, e.g., Carr et al. 2011; Freitag 2014; Marks et al. 2003; Po et al. 2003). Wessels et al. (2019, p. 188) provided an example of the Elgin Valley, South Africa, where pressures on water resulted from processes that were human-induced. This led to water pollution and the reuse of wastewater for irrigation being contested. Wessels et al. concluded that not only was the pollution itself contested, but also the meaning of the pollution, such as defining pollution problems, sources and responsibilities. Agreeing on solutions is an equally political process, despite being presented merely as a practical problem. As a result, complex situations arise, co-constituted by social, natural and political processes, and are reduced to mere biophysical problems that can be solved 'objectively', covering the often deeply political nature of the proposed solutions (Wessels et al. 2019, p. 188). Knowing water stems from the idea that many problems in water are complex. These complex problems are groups of interrelated problems, characterised by high levels of uncertainty, a diversity of competing values and a multitude of interest groups who may have different worldviews and different interpretative frameworks. The solution to these problems depends on how the problem is framed and vice versa; defining the problem depends on the envisioned solution (Zwarteveen 2015, p. 10).

A relevant example of how solutions are determined based on how problems are framed is illustrated in the book conducted in Carolina, Mpumalanga, South Africa, a predominantly coal-mining centre (Moeng 2018, p. 2622). According to the community's perceptions, poor water quality was due to improper mine closure, poor maintenance of water infrastructure and unknown sources. AMD was the leading cause of poor water quality, and despite community members possessing limited knowledge of AMD, they were still able to link their health-related problems to contaminated water. This meant that individual knowledge, views and experience of poor-quality water influence how the problem is framed and could have influenced the participants' responses. Respondents who agreed about the quality of water differed on the time frame relating to the quality of the water. Moeng (2018, p. 2636) deduced that perceptions of water quality are subjective and influenced by formal knowledge of what constitutes good-quality water.

The Carolina example demonstrated that water knowledge comes from somewhere (it is local), and understanding how knowledge travels enables an understanding of why it is viewed that way. At this point, it is necessary to ask why certain knowledge of water travels and carries greater authority than others (Zwarteveen 2015, p. 17). For instance, in the case of a mining operation, does it entail that the scientific measurements or changes in water qualities and quantities carry more weight in evaluating the environmental behaviour of the mining company, rather than people living with the changes brought about by the mining company? Zwarteveen indicated that the point was not to determine who was accurate (i.e. the mining company or the scientist) but that realities were accepted in specific historical, cultural and material practices, linking them to specific histories. This enables the transfer of knowledge; whose knowledge matters and counts, and on what grounds. Mpofu et al. (2018, p. 78) demonstrated that the complexity of such problems, such as water and mineral governance in South Africa, was attributed to highly politicised issues. These are questions about truth, as well as about how powers of authority exist and operate. Understanding that these different types of knowledge and knowers might never be accepted but, instead, will always be contested, meaning being realistic about the discussions and struggles about what knowing and dealing with water presents (Zwarteveen 2015, p. 21).

Strang (2004, p. 2) suggested that reasons for these different relationships with the use of water were because of self-interest, as described by policymakers and environmentalists. Her research indicated that patterns of use were manifested in social, economic and political relations, with close ties in power structures and control of resources between access to water and political empowerment. The political–ecological examination of hydrosocial processes reveals the inherently conflict-ridden nature of the process of socio-environmental change and teases out the inevitable conflicts that infuse socio-environmental change. Social relations will need particular attention by various users and managers of the water, respectively, through which hydrosocial transformations take place (Swyngedouw 2009, p. 58). Krause and Strang's (2016, p. 634) book on how social and hydrological relationships are interconnected and mutually constitutive suggests that better management and policy design will lead to a much deeper understanding of the role of water in humans' social lives. Hydrosocial networks or waterscape actors compete to establish a hydro-territorial order that is in line with their interests and beliefs. Threats to the hydrosocial territory structure stem from personal interests, as in the example of the Elgin Valley where agriculture was the main stakeholder (Wessels et al. 2019, p. 189). Wessels et al. suggested that studies from other sites in South Africa indicated that water struggles between commercial farmers and communities resulted from existing power relations at a local level.

How water management is embedded in power relations is best depicted in Strang's (2004) example of the Stour Valley in Dorset, United Kingdom. Water is constantly a representation of social, economic and political relationships, serving as an indicator of the extent to which identity, power and resources are shared. The Stour (a river in the United Kingdom) was "the community's source of life" (Strang 2004, p. 23). Water flowing in the Stour is both natural and cultural/social, resulting from changes in the spatial, temporal, physical and ideational landscape. Its material qualities are inherent and also resulted in a change in context, their emotional attachment to water remained for centuries, despite the new developments. People were not only physically linked to the river, but also through social terms, enabling them to share knowledge and values. Signifying water as social, 'a connective substance' to wealth, power and emotions expresses the relations people have with water. People still reflected on past communities and systems of management to voice ideas and construct their 'ideal' models (Strang 2004, pp. 245–246). These filter into decisions made about water use, and powerful influences in every disagreement about ownership, access and control of the resource.

The transition of the Stour into the adapted landscape demonstrated that the arrangement of water resource ownership and management was an extremely accurate indicator of social, political and economic relationships between groups, revealing their relative power and status and the influence of their particular values and beliefs. "In the historic patterns of enclosures in Dorset there is clearly a relationship between control over resources, social identity and the empowerment – or disempowerment – of particular groups" (Strang 2004, pp. 245–246). These underlying forces were revealed in the physical arrangements of the way in which water was managed. Further, people's emotional attachment to water, because it was placed in the hands of a specific manager, led water users to reduce management to the 'untrustworthy other'. The outcome is an ever-increasing gap between the institutions that now manage water resources at a national level and what remains of local control of water represented by riparian industry, conservation activities and grassroots political involvement (Strang 2004, p. 247).

Broadly, studies on the social construction of water show how engagement with water entails the material representation of particular ideas (Strang 2008). Bakker (2012, p. 617) referred to water's materiality, the role that its biophysical and ecological characteristics play in shaping human perceptions, broad constructions and responses to water. *Materiality* refers to an engagement with many aspects of water. Water is necessary for life and ecological health of deep spiritual significance which is highly sensitive to pollution, but necessary for industrialisation, urbanisation and agriculture. Bakker (2012, p. 621) suggested that the concept of materiality was an acknowledgement that the things, such as dams, that make a difference in the way social relations unfold are not merely pre-given substrates that enable the social action. They are historically and geographically produced in a way that is simultaneously socio-natural. There is immense competition between economic sectors for the use of water and its disposal, making water intensely political in a conventional sense; it is connected to contested relationships of power and authority (Bakker 2012, p. 616).

By exploring the power relations that reinforce nature–society relations, political ecologists have increased the understanding of water. In doing so, they have delivered important insights into some of the perspectives, institutions and processes of water governance, such as water privatisation, integrated water resource management, hydraulic engineering water technologies, hydrological studies and social struggles (Budds and Hinojosa 2012, p. 119). This indicated that social constructs embed power relations that suggest making the most of water use effectively in order to solve the 'global water crises'. Scientists' perceptions of which power struggles adequately manage water affect their definition of a crisis (Trottier 2008, p. 197). "Our knowledge is formed by looking at the world from some perspective or other and is in the service of some interests rather than others" (Burr 2003, p. 6).

Michel Foucault (cited in Bălan 2010) suggested that the issue with power was central to the relations between society, individuals, groups and institutions. Harrison and Dye (2008, p. 198), in their book on power and society, explained what distinguished government power from the power of other institutions, groups and individuals. The government's role is to maintain order and, thus, lay down the rules of the game in conflict and competition between individuals and institutions in society (Harrison and Dye 2008, p. 198). The ideal setting where power in action can be viewed is relations between the individual and society, especially in institutions. Foucault (cited in Bălan 2010) analysed power in terms of how "various institutions exert their power on groups of individuals and how the latter affirm their own identity and resistance to the effects of power". The aim is to examine how power takes place in daily interactions between people and institutions, and how the social constructions of water quality are linked to power relations. Foucault (cited in Bălan 2010) was less concerned with the oppressive aspect of power, and more with the resistance of those on whom the power was exercised. This book will explain how various elements or acts of power are linked to individuals' social constructions of water quality.

2.5 Conclusion

From a social science perspective, water is intrinsically social, and people construct and redefine their realities through social interactions. This book uses a social constructionist approach to explore how people (cognitively) frame water. This influences how people find solutions for water-related issues based on their interests. What is known about water and its quality is therefore socially constructed based on how water is framed, and how it is managed or governed. Water is highly politicised in South Africa, leading to solutions embedded in power relations and based on vested interests. Power is therefore central to relations between individuals, society and institutions. How people frame and value water indicates how they resist and challenge the power of the institutions responsible for water management due to their culture, knowledge, meanings and values of what they consider water and water quality. The next chapter provides a background to the study area by presenting a description of the East Rand and the eastern basin, and activities that take place in the area. A more detailed discussion on the governing bodies involved in the Blesbokspruit is presented as well as how the data were collected for this book.

References

Aguilera-Klink, F., Perez-Moriana, E. & Sanchez-Garcia, J. 2000. The social construction of scarity: The case of water in Tenerife (Canary Islands). *Ecological Economics*, 34(2):223–245. Available at: http://www.sciencedirect.com/science/article/pii/S0921800900001609. Accessed on: 8 March 2017.

Andrews, T. 2012. What is social constructionism? *Grounded Theory Review: An international Journal*, (1)11. Available at: http://groundedtheoryreview.com/2012/06/01/what-is-social-constructionism/. Accessed on: 6 March 2017.

Bălan, S. 2010. M. Foucault's view on power relations'. *Bucharest Academy of Economic Studies*, 11(61). Available at: https://www.researchgate.net/publication/321161337. Accessed on: 21 April 2020.

Bakker, K. 2003. The political ecology of water privatisation. *Journal Studies in Political Economy A Socialist Review*, 70(1):35–58. Available at: https://www.tandfonline.com/doi/abs/10.1080/07078552.2003.11827129?journalCode=rsor20. Accessed on: 20 May 2020.

Bakker, K. 2012. Water: Political, biopolitical, material. *Social Studies of Science*, 42(4):616–623.

Boelens, R. 2014. Cultural politics and the hydro-social cycle: Water, power and identity in the Andean Highlands. *Geoforum*, 57:234–247. Available at: http://www.sciencedirect.com/science/article/pii/S0016718513000432. Accessed on: 4 April 2017.

Bruffee, KA. 1986. Social construction, language, and the authority of knowledge: A bibliographical essay. *College English*, 48(8):773–790. Available at: http://www.jstor.org/stable/376723?seq=1#page_scan_tab_contents. Accessed on: 4 April 2017.

Budds, J. & Hinojosa, L. 2012. Restructuring and rescaling water governance in mining contexts: The co-production of waterscapes in Peru. *Water Alternatives*, 5(1):119–137.

Burr, V. 2003. *Social constructionism* (2nd edition). London & New York, NY: Routledge.

Burr, V. 2015. *What is social constructionism?* (3rd edition). London & New York: Routledge.

Carr, G., Potter, RB. & Nortcliff, S. 2011. Water reuse for irrigation in Jordan: Perceptions of water quality among farmers. *Agricultural Water Management*, 98:847–854.

Castree, N. & Braun, B. 1998. The construction of nature and the nature if construction: Analytical and political tools for building survivable futures. In *Remaking reality: Nature at them.* Edited by Castree, N. & Braun, B. New York, NY: Routledge:3–42. Available at: https://books.google.co.za/books?hl=en&lr=&id=zgKIAgAAQBAJ&oi=fnd&pg=PA2&dq=Castree+and+Braun+1998&ots=WOtC7pKQh&sig=AprMz_CRx75HXmpWFIICSA_Xi_g#v=onepage&q=Castree%20and%20Braun%201998&f=false. Accessed on: 3 April 2017.

Castree, N. 2001. Socializing nature: Theory, practice and politics. In *Social nature: Theory, practice and politics*. Edited by Castree, N. & Braun, B. Oxford and Malden: Basil Blackwell: 1–22.

Chen, C. 2013. Mapping water: Thinking with watery places. In *Thinking with water*. Edited by Chen, C., MacLeod, J., & Neimanis, AG. London: McGill-Queen's University Press: 274–299.

Demeritt, D. 1999. Contested natures. *Environment and Planning: Society & Space*, 17(3):372–374.

Demeritt, D. 2001. Being constructive about nature. In *Social nature: Theory, practice and politics*. Edited by Castree, N. & Braun, B. Malden, MA: Blackwell Publishers: 22–40.

Demeritt, D. 2002. What is the social construction of nature? A typology and sympathetic critique. *Progress in Human Geography*, 26(6):768–790. Available at: http://journals.sagepub.com/doi/abs/10.1191/0309132502ph402oa?journalCode=phgb. Accessed on: 8 March 2017.

Freitag, A. 2014. Naming, framing, and blaming: Exploring ways of knowing in the deceptively simple question 'what is water quality?' *Springer Science & Business Media*, 42(2):325–337. Available at: https://link.springer.com/article/10.1007/s10745-014-9649-5. Accessed on: 8 March 2017.

Gergen, MM. & Gergen, KJ. 2004. Reflections: Between Narcissus and Dorian Grey. *Community & Applied Psychology*, 14(4):299–301. Available at: http://onlinelibrary.wiley.com/doi/10.1002/casp.775/abstract. Accessed on: 3 April 2017.

Hacking, I. 1999. *The social construction of what?* London: Harvard University Press. Available at: https://books.google.co.za/books?hl=en&lr=&id=XkCR1p2YMRwC&oi=fnd&pg=PA1&dq=social+construction&ots=NF1cQ4Ysl3&sig=UmjKQz-xryP7rC9GMSROuIWazak#v=onepage&q=social%20construction&f=false. Accessed on: 6 March 2017.

Harrison, BC. & Dye, TR. 2008. *Power and society: An introduction to the social sciences* (11th edition). Boston, MA: Thomson Wadsworth.

Harrison, CM. & Burgess, J. 1994. Social constructions of nature: A case study of conflicts over the development of Rainham marshes. *Transactions of the Institute of British Geographers*, 19(3):291–310. Available at: http://www.jstor.org/stable/622324?seq=1#page_scan_tab_contents. Accessed on: 3 April 2017.

Heft, H. 1981. An examination of constructivist and Gibsonian approaches to environmental psychology. *Springer*, 4(4):227–245. Available at: https://link.springer.com/article/10.1007/BF01375628. Accessed on: 6 March 2017. https://www.jstor.org/stable/2393414?seq=1. Accessed on: 10 October 2018.

Ibarra, H. & Andrews, SB. 1993. Power, social influence, and sense making: Effects of network centrality and proximity on employee perception. Administrative Science Quarterly, 38(2):277–303. Available at: https://www.jstor.org/stable/2393414?seq=1. Accessed on: 10 October 2018.

Kaika, M. 2003. Constructing scarcity and sensationalising water politics: 170 days that shook Athens. *Antipode*, 35(5):919–954.

Krause, F. & Strang, V. 2016. Thinking relationships through water. *Society & Natural Resources*, 29(6):633–638.

Lien, SG. 2006. *Commodification of nature: Social construction of bottled spring water*. Oslo: University of Oslo. Available at: https://www.duo.uio.no/bitstream/handle/10852/17790/Imsdal.pdf?sequence=1. Accessed on: 1 November 2017.

Linton, J. 2010. *What is water? The history of a modern abstraction*. Vancouver: UBC Press.

Linton, J. & Budds, J. 2014. The hydrosocial cycle: Defining and mobilising a relational-dialectical approach to water. *Geoforum*, 57:170–180.

Marks, J., Cromar, N., Fallowfield, H. & Oemcke, D. 2003. Community experience and perceptions of water reuse. Available at: http://ws.iwaponline.com/content/3/3/9. Accessed on: 26 January 2017.

Mehta, L. 2003. Contexts and constructions of scarcity. *Economic and Political Weekly,* 38(48): 1–12. Available at: https://www.researchgate.net/profile/Lyla_Mehta/publication/251901079_Contexts_and_constructions_of_water_scarcity/links/560e4c8908ae967420111f88.pdf. Accessed on: 8 March 2017.

Mehta, L. 2011. The social construction of scarcity: The case of water in western India. In *Global political ecology.* Edited by Peet, R., Robbins, P. & Watts, MJ. London & New York: Routledge: 654–663.

Moeng, K. 2018. Community perceptions on the health risks of acid mine drainage: The environmental justice struggles of communities near mining fields. *Environment, Development and Sustainability*, 21:2619–2640. Available at: https://0-doi-org.oasis.unisa.ac.za/10.1007/s10668-018-0149-4. Accessed on. 1 June 2020.

Molle, F., Wester, P, Hirsch, P., Jensen, JR., Murray-Rust, H., Paranjpye, V., Pollard, S & Van der Zaag, P. 2007. River basin development and management. In *Water for life: A comprehensive assessment of water management in agriculture.* Edited by Molden, D. London; Earthscan: 585-624. Available at: https://www.researchgate.net/publication/40794254_River_basin_development_and_management. Accessed on: 5 September 2019.

Mpofu, C., Morodi, TJ. & Hattingh, JP. 2018. Governance and socio-political issues in management of acid mine drainage in South Africa. Water Policy, 20:77–89. Available at: https://iwaponline.com/wp/article/20/1/77/38136/Governance-and-socio-politicalissues-in. Accessed on: 1 June 2020.

Norman, ES. & Bakker, K. 2009. Transgressing scales: Water governance across the Canada–U.S. borderland. *Annals of the Association of American*, 99(1):99–117. Available at: https://www.tandfonline.com/doi/abs/10.1111/0004-5608.00157. Accessed on: 2 February 2017.

Ohlsson, L. & Turton, A. 2000. *The turning of a screw: Social resource scarcity as a bottle-neck in adaption to water scarcity*. Available at: https://www.soas.ac.uk/water/publications/papers/file38362.pdf. Accessed on: 24 May 2018.

Ohlsson, L. & Turton, A. 2003. *Water scarcity and social stability: Towards a deeper understanding of the key concepts needed to manage water scarcity in developing countries*. Available at: https://www.soas.ac.uk/water/publications/papers/file38360.pdf. Accessed on: 31 May 2018.

Oktem, O. 2015. *Water politics and political culture: Turkey's compatibility with the European Union*. London: Springer.

Palomino-Schalscha. M., Leaman-Constanzo. C. & Bond. S. 2016. Contested water, contests development: Unpacking the hydro-social cycle of the Nuble River, Chile. *Third World Quarterly*, 37(5):883–901.

Pickering, E. 2014. *The social construction of water in Dominica and how it has influenced use and exportation*. Phoenix: University of Arizona. Available at: http://media.proquest.com/media/pq/classic/doc/3570237761/fmt/ai/rep/NPDF?_s=%2FP0hpYCMw19x%2FWxRHTCaNR5X6xY%3D. Accessed on: 22 February 2017.

Pinch, T. & Bijker, W. 1987. The social construction of facts and artifacts: Or how the sociology of science and the sociology of technology might benefit each other. In *The social construction of technological systems: New directions in the sociology and history of technology*. Edited by Bijker W., Hughes T. & Pinch T. Cambridge, MA: MIT Press: 399–441.

Po, M., Kaercher, JD. & Nancarrow, BE. 2003. *Literature review of factors influencing public perceptions of water reuse* (CSIRO Land and Water, Technical Report 54/03). Available at: https://www.clearwater.asn.au/user-data/research-projects/swf-files/16-laying-the-foundation-for-confident-barrier-free-water-conservation-and-reuse-literature-review.pdf. Accessed on: 26 January 2017.

Schmidt, JJ. 2014. Historicising the hydrosocial cycle. *Water Alternatives*, 7(1):220–234.

Strang, V. 2004. *The meaning of water*. London: Berg.

Strang, V. 2008. The social construction of water. In Handbook of landscape archaeology. Edited by David, B. & Thomas, J.: 123-130. Available at: https://books.google.co.za/books?hl=en&lr=&id=9bRJDAAAQBAJ&oi=fnd&pg=PA123&dq=veronica+strang+2008&ots=EM7K-XIkwx&sig=fSQObJJmH0pEVhnr5jjv-Vw18fg#v=onepage&q&f=false.

Swyngedouw, E. 1999. Modernity and hybridity: Nature, Regeneracionismo, and the production of the Spanish waterscape, 1890–1930. Annals of the Association of American Geographers, 89(3):443–465. Available at: https://www.tandfonline.com/doi/abs/10.1111/0004-5608.00157. Accessed on: 2 February 2017.

Swyngedouw, E. 2009. The political economy and political ecology of the hydrosocial cycle. *Journal of Contemporary Water Research and Education*, 142:56–60.

Torregrosa, ML., de las Mercedes, D., Acosta, A. & Kloster, K. 2019. Water and society: The multidimensionality of water quality. In *Water quality in the Americas: Risks and opportunities*. Edited by The Inter-American Network of Academies of Sciences. Mexico: INAS: 20–26.

Trottier, J. 2008. Water crises: Political construction or physical reality? *Contemporary Politics*, 14(2):197–214.

Tvedt, T. 2015. *Water and society: Changing perceptions of societal and historical development*. London, New York: IB Tauris.

Wagemans, JP. 1986. Direct theory of perception: An evaluation by representatives of indirect theories of perception. *L'annee psychologique*, 86(2):261–273. Available at: http://www.persee.fr/doc/psy_0003-5033_1986_num_86_2_29144. Accessed on: 6 March 2017.

Wessels, M., Veldwisch GJ., Kujawa, K. & Delcarme, B. 2019. Upsetting the apple cart? Export fruit production, water pollution and social unrest in the Elgin Valley, South Africa. *Water International*, 44(2):188–205.

Whatmore, S. & Boucher, S. 1993. Bargaining with nature: The discourse and practice of environmental planning gain. Transactions of the Institute of British Geographers, 8(3):166–178. Available at: http://www.jstor.org/stable/622360?seq=1#page_scan_tab_contents. Accessed on: 3 April 2017.

Zwarteveen, M. 2015. *Regulating water, ordering society: Practices and politics of water governance*. Amsterdam: UNESCO Institute for Water Education, Universiteit van Amstar.

Chapter 3
Background on the Eastern Basin and the Blesbokspruit

Abstract This chapter provides background on the eastern basin with special focus on the significance of the Blesbokspruit. It describes the East Rand in the Gauteng Province. The intricate nature of the Blesbokspruit is presented and how dependent various stakeholders who reside or work in the eastern basin are on this watercourse. The Blesbokspruit wetland was named a Ramsar site in 1986 and it underwent extensive transformation due to industrial development over the years and the resultant ecological damage, which led to it being placed on the Montreux Record. The chapter then sets out the non-statutory bodies contributing to the governance of the Blesbokspruit by explaining the role of the Blesbokspruit Forum and Blesbokspruit Trust, which include members of various spheres of government that are responsible for managing the Blesbokspruit. A brief section on the research methodology is included.

Keywords East Rand · Eastern basin · Blesbokspruit wetland

3.1 Introduction

This chapter provides background on the eastern basin with a special focus on the significance of the Blesbokspruit. It begins by describing the East Rand in the Gauteng Province. The East Rand was known for its abundance of mineral resources and had become highly lucrative since the 1940s, which brought many people to the area. However, the many profitable years of mining had a detrimental role on the environment of the eastern basin and specifically the Blesbokspruit. The intricate nature of the Blesbokspruit is presented as well as how dependent various stakeholders who reside or work in the eastern basin are on this watercourse. The Blesbokspruit wetland was named Ramsar site in 1986, and it underwent an extensive transformation due to industrial development over the years and the resultant ecological damage, which led to it being placed on the Montreux Record (explained in Sect. 3.4). The chapter then sets out the non-statutory bodies contributing to the governance of the Blesbokspruit by explaining the role of the Blesbokspruit Forum

S. Naidoo, *Social Constructions of Water Quality in South Africa*,
https://doi.org/10.1007/978-3-030-98237-9_3

and Blesbokspruit Trust, which include members of various spheres of government that are responsible for managing the Blesbokspruit. A brief section on the research methodology is included.

3.2 The East Rand

The wider Witwatersrand, of which the East Rand forms part, is the most densely populated region in South Africa. This area has a Highveld summer rainfall that occurs between November and April, with the average annual rainfall varying between 650 mm and 950 mm (Hawley and Desmet 2020, p. 10). The East Rand is prone to extremely low temperatures during the winter season but has hot summers. The East Rand (Fig. 3.1) includes the towns of Boksburg, Brakpan, Benoni, Springs and Nigel, which all fall within the jurisdiction of the City of Ekurhuleni (CoE) in the Gauteng Province (Digby Wells Environmental 2015, p. ii). This book is mostly centred on Springs, where the eastern basin acid mine drainage (AMD) treatment plant is situated. The CoE covers an area of approximately 1975 km^2 (Hawley and Desmet 2020, p. 1), and the surface water in the CoE is made up of several dams (e.g. Cowles and Nigel dams) and streams, which include the Blesbokspruit.

Many of the towns on the East Rand, such as Springs, came into existence through mining activities. Some of the black townships in the CoE, such as Daveyton, Kwa-Thema, Duduza and Tsakane, were established to create a residence for the mineworkers (Labuschagne 2015, p. 11). The increased number of people residing in the area is due to the employment opportunities that mining created. However, this also led to social risks such as the insufficient supply of clean drinking water and safe sanitation. It sounds contradictory that Ekurhuleni, translated as a 'Place of Peace',[1] has environmental problems that mostly stemmed from people's excessive need for, and use of, water.

The East Rand is home to numerous important agricultural areas scattered across the region, and agriculture is known as the sector that requires the largest volumes of water (Environomics 2014, p. 18). The largest use of land (Fig. 3.2) is in the built-up urban group, accounting for 37%, cultivation agriculture covering 14% of the CoE, and mining accounts for 3% of land use (Hawley and Desmet 2020, pp. 20–21). Owing to growing business, the East Rand is described as a sanctuary for agricultural and mining activities and is proven to be unequivocal in the development and economic growth of South Africa. The East Rand is largely modified by urban, mining and agricultural development, and the CoE still supports threatened biodiversity and important ecological infrastructure within the grassland biome, which offers a variety of ecosystem services (Hawley and Desmet 2020, p. 1). Historical mining activity created a few lakes on the East Rand – such as Benoni, Boksburg and Germiston – which were used to supply the mines with water, before the Vaal Dam

[1] The Tsonga word 'Ekurhuleni' means 'Place of Peace' (South African Cities Network 2019).

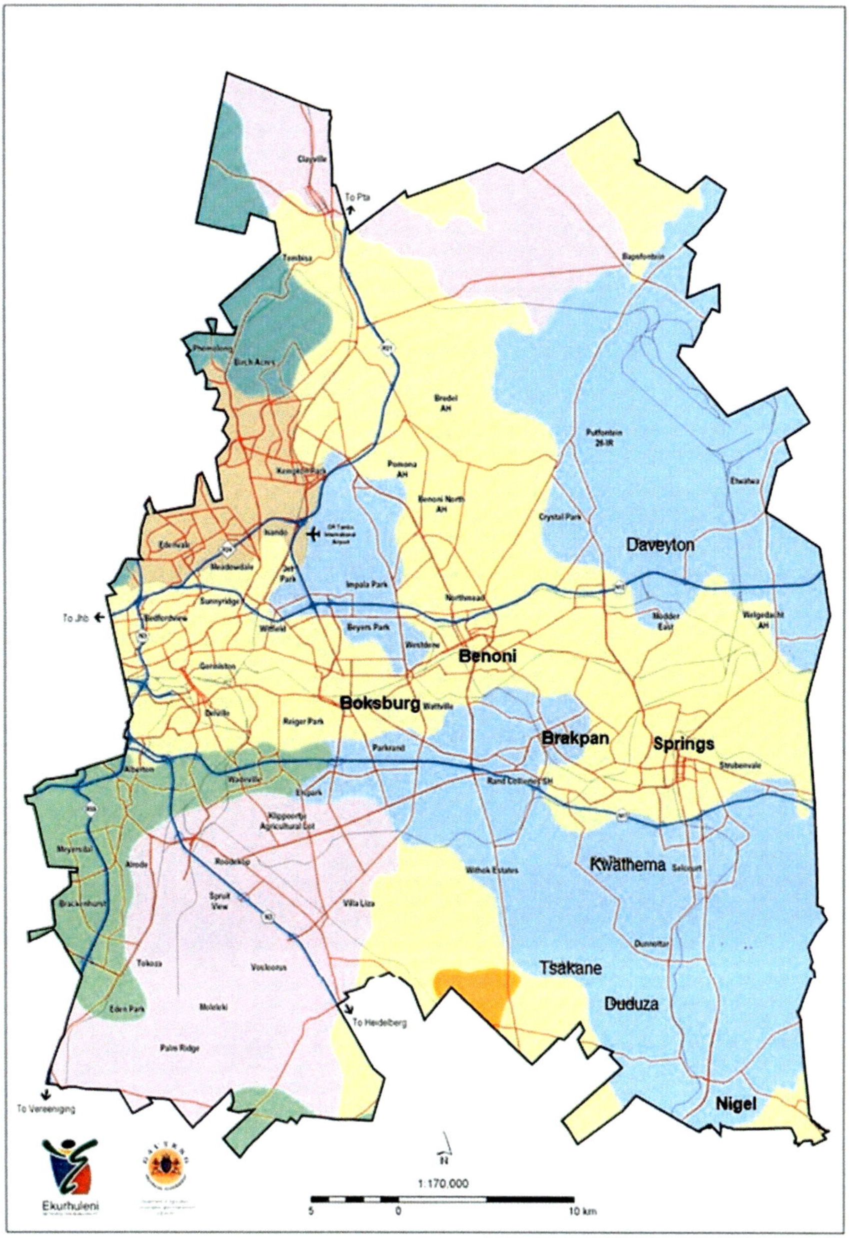

Fig. 3.1 Map of the East Rand. (EMM 2007b, p. 7)

was developed, and these lakes now serve recreational purposes (Labuschagne 2015, p. 1). These surface water areas act as sanctuaries to a substantial amount of South Africa's biodiversity and are threatened by grassland biomes (Labuschagne

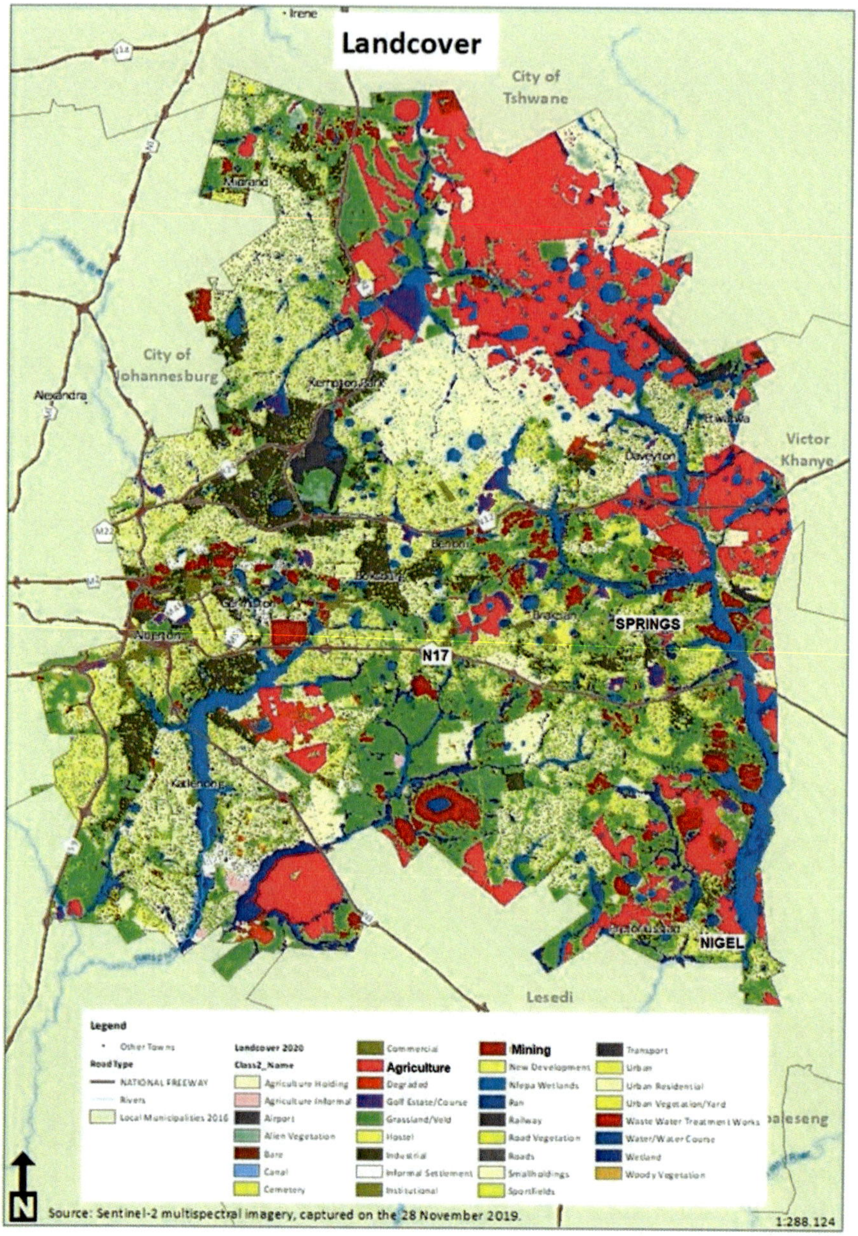

Fig. 3.2 Land use on the East Rand. (Hawley and Desmet 2020, p. 22)

2015, p. 1). However, because of the substantial degree of alteration of the natural landscape, most of the remaining ecosystems on the East Rand are threatened. These threats arise from the fact that the various uses of water for human purposes, such as agriculture, industry or for urban use, exceed the amount of renewable water available (Molle et al. 2007, p. 585). Molle et al. (2007, p. 585) suggest that this leads to over-commitment of the water resource due to users not accounting for environmental water requirements, lack of hydrological knowledge, uncertain water rights or poor governance, resulting in more water being used than the system (environment) can allow.

Labuschagne (2015, p. 1) notes that South Africa is known as a treasure house due to its valuable minerals. Its comparative advantage in terms of mineral endowment has not translated into an economic competitive advantage due to numerous challenges facing the mining industry. The industry is continually working on addressing these challenges to maintain and reserve its space in the global market, while addressing national and community needs (Neingo and Tholana 2016, p. 283). However, as illustrated in Fig. 3.2, there is little available land around the Blesbokspruit system to accommodate new economic development, especially since there are a few areas that have little or no impact already on the East Rand. There is a "significant challenge to densify and redevelop existing built-up areas, as well as old mining and other land that is no longer used optimally" (EMM 2007a, p. 5).

For the sake of clarity, it is important to explain the terminology used in this book, and therefore to distinguish between the East Rand, the eastern basin, the Blesbokspruit Wetland and the Blesbokspruit catchment. The East Rand is a group of towns, which fall under the CoE's jurisdiction. The East Rand can be differentiated from the eastern basin (see Fig. 3.3); a *basin* refers specifically to a river

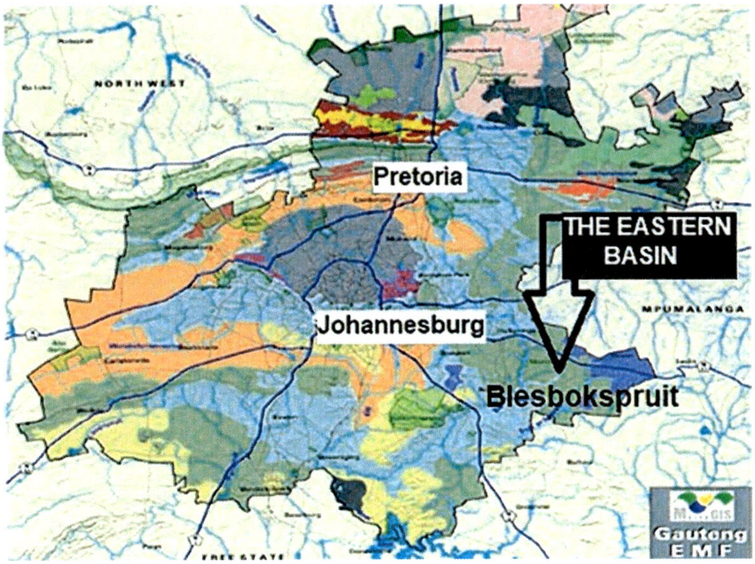

Fig. 3.3 Location of the Blesbokspruit and the eastern basin in the Gauteng Province. (Environomics 2014, p. 11)

drainage system, an area of land drained by a river or its tributaries (Milwaukee Riverkeeper 2015). Mining resources were found in this basin. Thus, mining overlaps with the geographical area, indicating that there can be residential areas situated where mining activity is taking place, as in the case of the eastern basin.

The Blesbokspruit catchment forms part of the eastern basin, which is on the East Rand, within the Upper Vaal Water Management Area, which is one of the water management areas that have been identified in the National Water Resource Strategy of South Africa (du Plessis et al. 2014, p. 2950). The Blesbokspruit catchment (see Fig. 3.4) drains an area of almost 1000 km² and functions within the Vaal

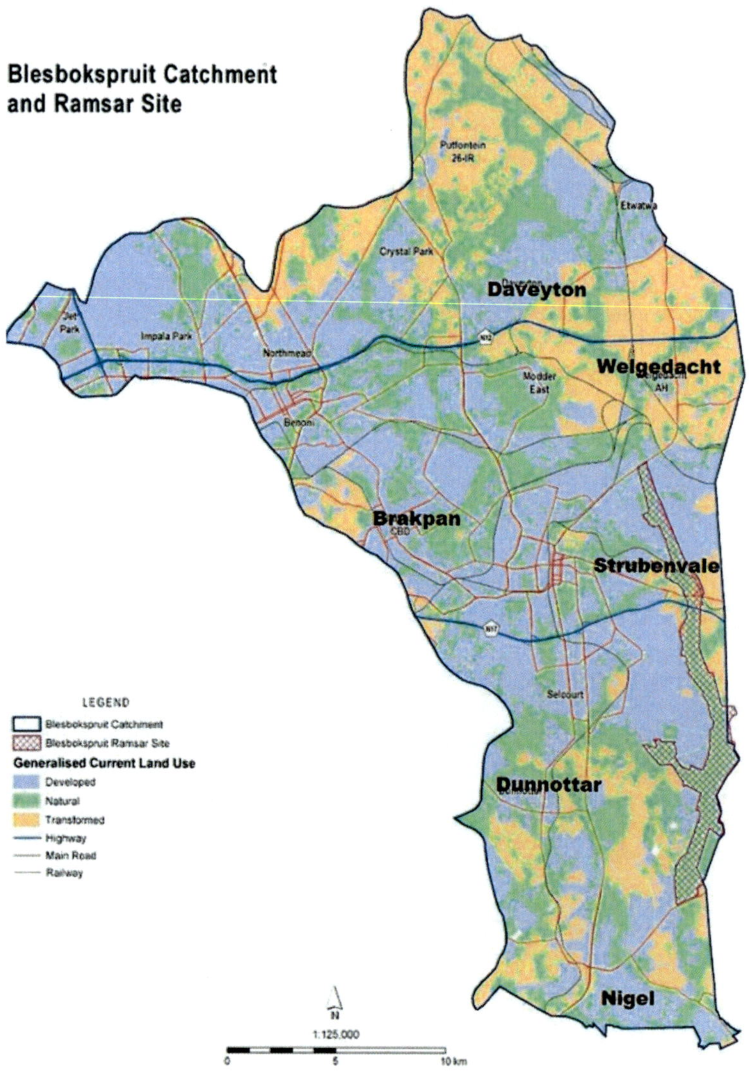

Fig. 3.4 Blesbokspruit catchment. (EMM 2007b, p. 18)

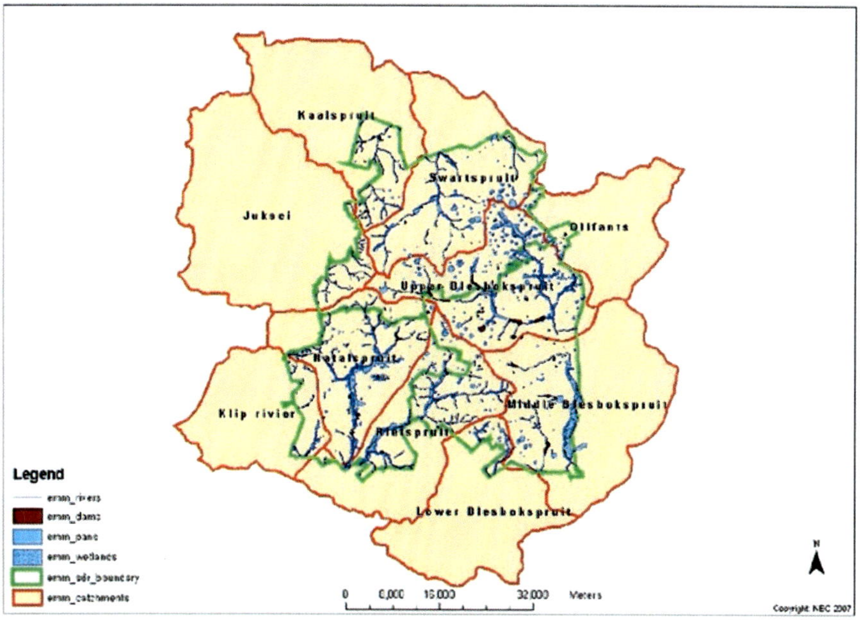

Fig. 3.5 Catchment management areas in the City of Ekurhuleni. (EMM 2007b, p.14)

Barrage and Vaal Dam catchments of the Upper Vaal Water Management Area. The Blesbokspruit (see Fig. 3.5) is managed by a catchment management agency, which aims to achieve effective communication with all stakeholders in the water management area to formulate and implement a catchment management strategy. Catchment management is achieved through the Blesbokspruit Forum (discussed in Sect. 3.5.1) (Blesbokspruit Forum Charter 2003, p. 2; RSA 1998). The Blesbokspruit is the watercourse that is the focus of this book. For the purpose of the book, the eastern basin is referred to because of the relevance of mining and its impact on the water quality of the Blesbokspruit. Depending on the context, the term Blesbokspruit catchment is used when referring to the management of the water, and the term Blesbokspruit wetland or river is used when referring to the uses of the water.

3.3 The Eastern Basin

The *eastern basin* – also referred to as the East Rand basin (see Fig. 3.6) – refers to the mining-related areas to the east of Johannesburg, and the term is used in the context of dealing with AMD on the Witwatersrand. This terminology is based on the three underground mining basins, namely (1) eastern, (2) western and (3) central established on the Witwatersrand after the discovery of gold in 1886 (Adler et al. 2007, p. 34). The discovery of gold is attributed to George Walker and George

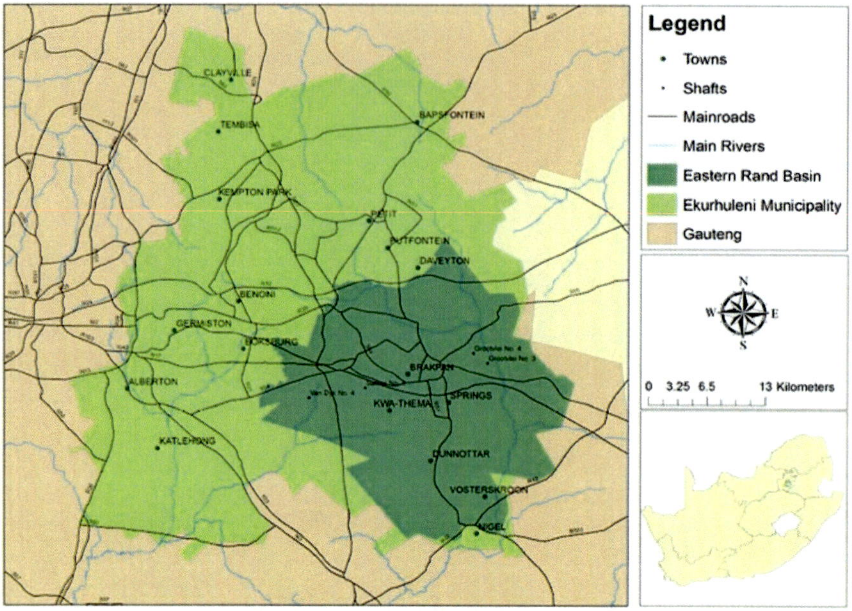

Fig. 3.6 The eastern basin in the Gauteng Province. (Labuschagne 2015, p. 2)

Harrison. They identified "surface outcrops of gold-rich conglomerate on old farmland", called *Langlaagte*, near the centre of Johannesburg (Kirk et al. 2004, p. 534). After this discovery, the "outcrop of the Main Reef Group, which hosts the most important gold-bearing conglomerate reefs of the Witwatersrand, was soon traced along the Central Rand for some 40 kilometers (km), from the Durban Roodepoort Deep mine in the west to the East Rand Proprietary Mines in the east" (Tucker et al. 2016, p. 106). This is how gold in the west, east and central basins were discovered. The Witwatersrand basin is known as the largest goldfield in the world, having produced half of the gold ever mined (Kirk et al. 2004, p. 534); in total over 52,000 t of gold, with approximately 30,000 t still available to mine (Tucker et al. 2016, p. 106; Hartnady 2009, p. 328). Partly owing to gold-mining activities on the Witwatersrand, including the East Rand, in 2009 South Africa dominated the world as the number 1 gold producer, indicating the extent of mining that took place. South Africa is currently the fifth largest producer of gold, with China, Australia, Russia and the United States having overtaken the country's position (Neingo and Tholana 2016, p. 283).

The eastern basin is about 30 km long and 20 km wide and has a mine lease area that extends for 768 km^2 (Fig. 3.7) (Labuschagne 2015, p. 1). Gold mining has occurred in the eastern basin since the discovery of gold in 1886 and reached its peak in the 1940s and 1950s. In 1955, there were 24 gold mines and 90 shafts existed, but due to the fixed gold price and high working costs, there were several mine closures between 1950 and 1960, with mines being made inactive, waiting for closure certificates or completely abandoned (Labuschagne 2015, p. 12; Scott 1995, p. 19). Extensive coal mining in the eastern basin also took place. Coal mines began

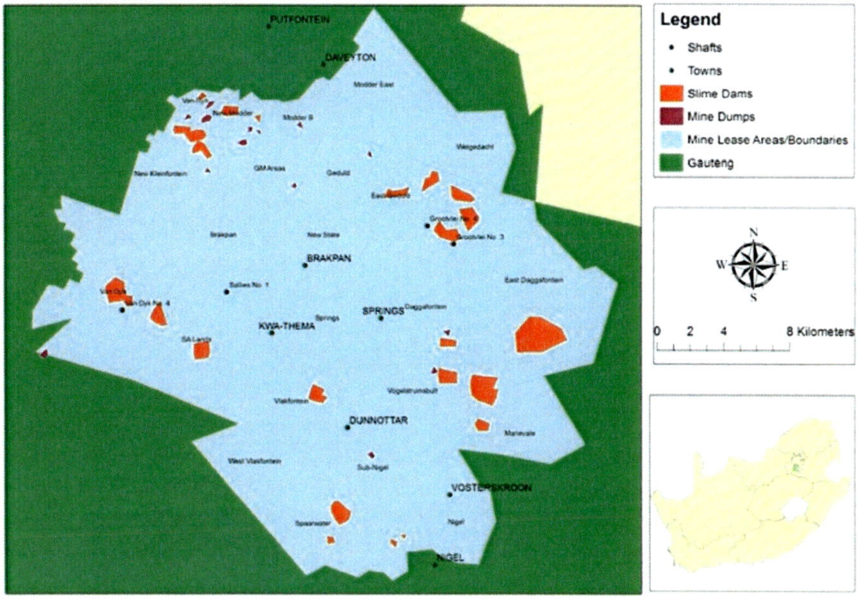

Fig. 3.7 Eastern basin mine lease areas. (Labuschagne 2015)

operating from the early 1890s and ended in the late 1940s. However, the economic resource potential still existed in the area, which led to active mining for many more years. As a result, even though most gold mining in this basin had already reached its lifespan and began to decline, during its peak it had 28 mines and produced about 10,000 tons of gold (Labuschagne 2015, p. 2). Thousands of people earned their living off mining during its predominant years, implying that "the mines of the Witwatersrand have shaped the economy of this country" (Scott 1995, p. 6) (see Fig. 3.8 for how widespread mines are across the Gauteng Province). However, by 2011 most of the mines were already closed, and damage caused began to surface increasingly (Ambani 2013, p. 88). The damage was due to "depth, geological complications and their primitive, poorly planned, beginnings" (Scott 1995, p.1). The mining industry had underestimated the potential life of the gold mines of the Witwatersrand (Tempelhoff et al. 2007, p. 107). Figure 3.9 illustrates the rate at which mining excelled and then declined in the eastern basin, as well as how many leading mining companies operated at a time.

Owing to the fact that mineral recovery is not a renewable process, once a mine reaches the end of viable economic production, the mine needs to be closed (Milaras et al. 2014, p. 1). Sustainable mine closure must be in place not to negatively impact the area that was mined so that the land can be used in future. This requires thorough and efficient management of mine closure, which creates "significant management problems" for governments, communities and mining companies (Milaras et al. 2014, p. 2). This implies that mining has a stark effect on the natural environment, making it impossible to measure the benefits versus the obstacles. Mining began

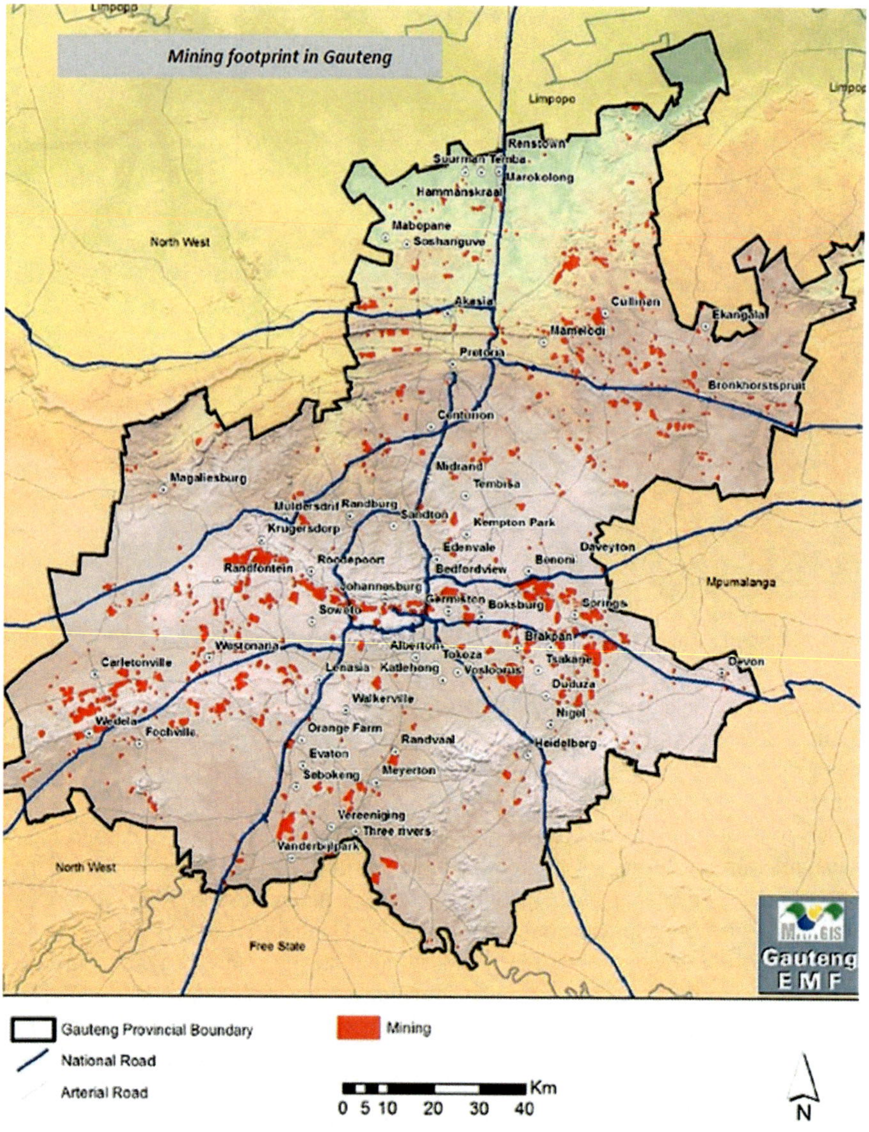

Fig. 3.8 Mining in Gauteng Province. (Environomics 2014, p. 50)

over a century ago, and its ongoing detrimental impact on the environment was possibly never anticipated when gold was discovered. Similarly, when gold was discovered, the process of extracting the resource had to be learnt, and during this process the impact it could pose to the environment was yet to be revealed. Therefore, ways to address the impact also had to be tried and tested. Consequently, the commencement of mining led to the discovery of mining's destruction of the environment.

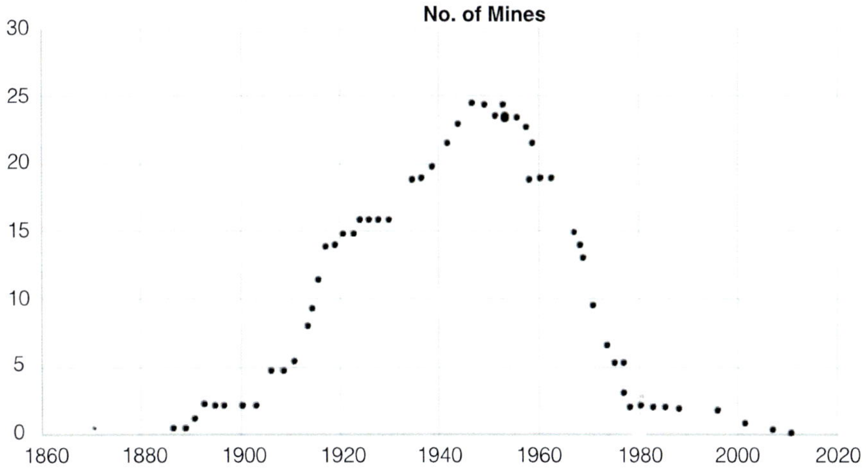

Fig. 3.9 Mining companies in the eastern basin since the late 1880s. (Adapted from Scott 1995, p. 27; Ambani 2013, pp. 87–88)

As gold mining developed on the East Rand, the underground operations became interlinked and fewer mines used dewatering (to remove excess water). Underground mining results in the collapse of the overlying rock strata, and when mining terminates, the voids in the fractured rock become filled with water and decanting occurs from the lowest opening. Further, opencast coal mining involves blasting and removal of the rocks overlying the coal layer, which is removed completely. Opencast coal mining is known to destroy the natural groundwater systems and severely alter the nature of groundwater and surface water interactions because decanting takes place almost a decade or more after mining ceases (Environomics 2014, p. 28).

Environmental problems stemmed from ongoing groundwater control, water resource contamination by acid mine drainage and other factors of pollution caused by illegal underground mining by Zama Zamas[2], "the glory days of South African gold mining appear to have arrived finally at a shameful end" (Hartnady 2009, p. 329). The commencement of environmental degradation associated with the formerly booming industry now posed a severe threat and is costing far more than the value of some of the gold ever mined (Hartnady 2009, p. 329). Further, coal is increasingly becoming available in the eastern basin and extensive coal mining operations occur (De Jager, personal interview, 2018; Madden, personal interview, 2018; Pillay, personal interview, 2018; Storey, personal interview and site visit, 2018; Van der Merwe, personal interview, 2018).

Coal mining is vital because through the process resources are obtained that cannot be produced through farming or cannot be made in a factory, such as electricity, which is imperative considering South Africa's struggle with ongoing power cuts

[2] This is a term used in South Africa for an illegal miner.

due to demands for electricity (Creamer 2019a). Until alternative more sustainable energy sources can replace the need for coal mining, the coal industry is crucial for human sustainability, and vital for economic growth and social upliftment (Environomics 2014, p. 54). While this remains the case, enormous cost implications and damage to the environment occur. Coal mining activities include the creation of slimes dams, quarries, mine dumps and disturbed land (Labuschagne 2015, p. 10), and the continuous pumping to remove inflow water.

The aforementioned make it critical for new technology to be experimented with globally as part of an initiative to ensure mining is safer and more efficient. New initiatives for safer mining are jointly funded by the private and public sectors. Representatives are utilised to "provide longevity to the declining South African mining industry" (Creamer 2019b). The main aim is "to mark the end of mining's fragmented past" and therefore to advance in safer, healthier and more efficient mining to take mining to a level for the benefit of the South African people (Creamer 2019b).

The Blesbokspruit experienced declining water quality over the years in which mining took place and more so when mining ceased in 2011, which implied that pumping of water had come to an end (discussed in Sect. 4.3). In addition, as the population grows, increasing volumes of water from wastewater treatment plants are discharged into the Blesbokspruit, further exacerbating an already sensitive system.

3.4 The Blesbokspruit and Its Wetland

The Blesbokspruit area of study includes Springs (see Figs. 3.10 and 3.11), where the AMD treatment plant is situated. Springs[3] includes several residential areas, such as Strubenvale, Casseldale, Grootvaly Agricultural Holdings, Bakerton, Welgedacht and Aston Lake. The town of Nigel is downstream of the Blesbokspruit and community members from this town were also interviewed for this book.

There are 75,300 individuals and 23,700 households that fall within the Blesbokspruit area (Digby Wells Environmental 2015, p. 82). The potential economically active percentage of the population is 72%, and 6% are of retirement age. Of the households, 9% reside in rural areas, including farms and smallholdings, and 92% access water through municipal means or other water service providers. The majority of the area is considered established as most houses have been in existence for over 20 years (Digby Wells Environmental 2015, p. 83). The economic development in the area was discussed in Sect. 3.2.

The book identified stakeholders and key individuals who interact with the Blesbokspruit to explore how various social constructions of the water quality are formed, and why they differ. Stakeholders who interact with the Blesbokspruit

[3] This study refers to the Springs community unless otherwise stated.

Fig. 3.10 The Blesbokspruit. (Photographs by researcher)

include environmental activists, communities (mainly residential property owners from Springs and Nigel), researchers, government (local, provincial and national), industry (mining and wastewater treatment works), tourism (Marievale Bird Sanctuary, and wedding and conference venues) and commercial agriculture (including feedlots). The satellite image in Fig. 3.12 indicates where these stakeholders are situated in proximity to the Blesbokspruit and, therefore, users of the water. Residential areas are situated alongside the river and irrigation points are visible. Recreational activities along the Blesbokspruit happen at places such as the Stable Inn Lodge and Conference Venue, and the Riverside Country Estate wedding venue. Marievale Bird Sanctuary, the Suikerbosrand Nature Reserve and the Karan Beef Feedlot are situated further downstream. The eastern basin AMD treatment plant is situated upstream, on the Grootvlei Mine site. The AMD treatment plant discharges from 80 Mℓ up to 110 Mℓ of water per day into the Blesbokspruit. This created increased volumes of water flowing into the Blesbokspruit, which led to the flooding of the banks of property situated alongside the river.

The Blesbokspruit originates to the north of Benoni and Daveyton and flows southwards through Springs, Heidelberg and Nigel towards the Vaal River (EMM 2007a, p. 10; Springs Advertiser 2014a). Most of the Blesbokspruit catchment though falls within Nigel. Half of the Blesbokspruit wetland is protected by the Marievale Bird Sanctuary, and the other half is owned by the Anglo-American Group, due to being situated on land they own (Thorius 2004, p. 18). For clarity, the Grootvaly Wetland Reserve lies off Welgedacht Road and is a high-altitude wetland

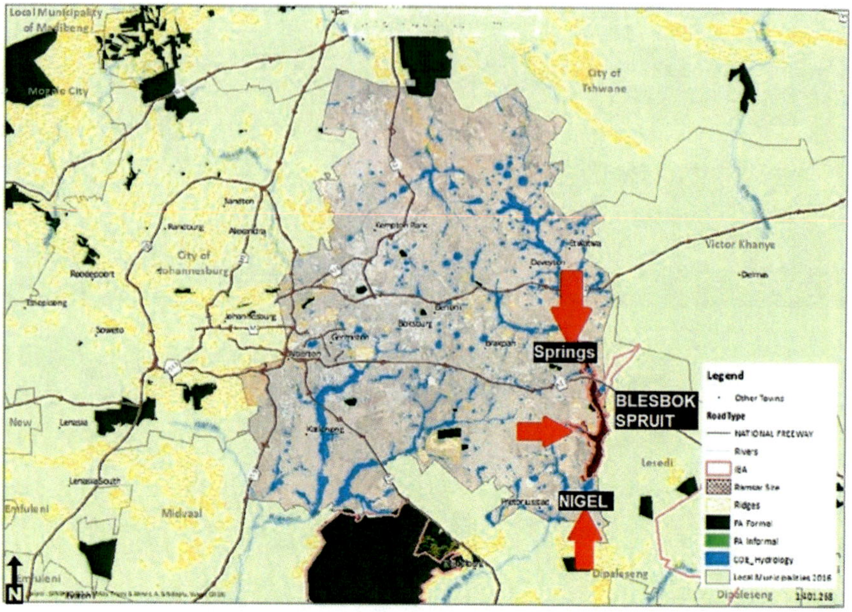

Fig. 3.11 Location of the research area. (Hawley and Desmet 2020, p. 24)

that runs along the Blesbokspruit River. The Grootvaly wetland lies to the north of the Blesbokspruit and at the southern end is the Marievale Bird Sanctuary; and these two form the Blesbokspruit Wetland Reserve (Ambani 2013, p. 63; South Africa Venues 2019). In this book, I refer to this as the Blesbokspruit wetland.

The Blesbokspruit wetland is one of the largest wetlands on the Highveld and was declared a Ramsar site in December 1986 (see Fig. 3.13), meaning it gained the status of international significance due to its unique ecosystem (Krige 2018; Madden, personal interview 2018). The Ramsar Convention[4] on Wetlands of International Importance was agreed to in Iran as one of the first international environmental treaties to monitor wetland degradation, and therefore to protect these ecosystems as habitats for wildlife and birds (Ambani 2013, p. 17). Wetlands that are classified on this list are those requiring special attention due to their significant value.

The Blesbokspruit (see Fig. 3.14) is one of the most important river systems in Gauteng and forms part of the Vaal River Catchment, along with the Klip River and the Natal Spruit. The Blesbokspruit flows into the Suikerbos River where it then joins the Klip River (Labuschagne 2015, p. 8). The Blesbokspruit, covering an area of 1427 km^2, is one of two main streams in the East Rand, the Rietspruit (which is

[4] South Africa was one of the first signatories to the Ramsar Convention in 1975 (Ramsar website 2017).

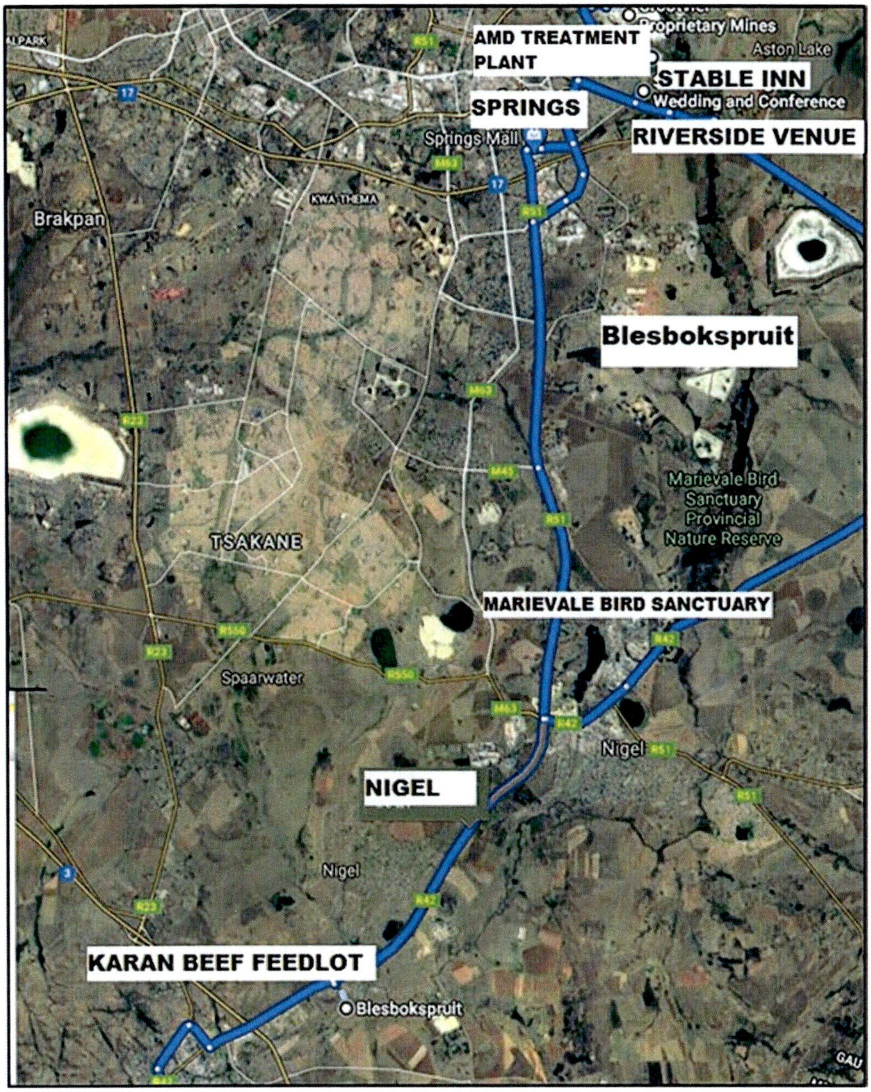

Fig. 3.12 Stakeholders who interact with the Blesbokspruit. (Google Maps 2020)

820 km^2) being the other (Scott 1995, p. 16). It extends over 22 km between the R22 and R555 road routes and is about 7 kilometres wide (Springs Advertiser 2014a).

The Marievale Bird Sanctuary is situated within the upper reaches of the river. It is known as one of the most popular attractions in Gauteng and covers over 1000 ha of the Ramsar site (Gauteng Tourism Authority 2020; Madden, personal interview, 2018). According to Hawley and Desmet (2020, p. 19), its ecosystem provides protection for over 450 bird species, including water birds (Gauteng Tourism Authority 2020). Approximately 18 species are of conservation concern and are reliant on the

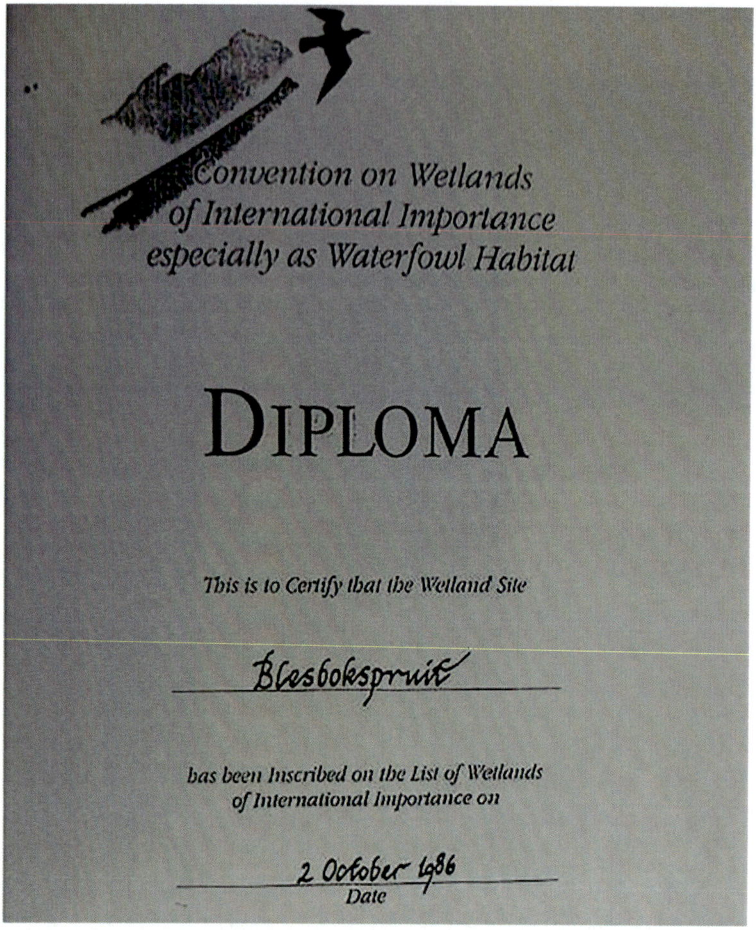

Fig. 3.13 Certificate indicating the recognition of the Blesbokspruit wetland as a Ramsar site. (Photograph by Stan Madden)

habitat in the CoE, 2 of these species are endangered, 9 are vulnerable as they are reliant on wetland habitats, and 7 are threatened (Hawley and Desmet 2020, p. 19). In 2016, 3082 birds were counted, according to the Coordinated Wetland Aquatic Count (Krige 2018). To the east of the wetland there are extensive natural wetlands, and the area to the west is highly developed by agriculture and human settlements, mines, waste disposal sites and wastewater treatment works, all impacting negatively on the quality of the water of the Blesbokspruit (EMM 2007a, p. 10). Further, most of the Blesbokspruit is filled with reeds, proven to be necessary for the system, as it absorbs chemicals and other substances from the water into their roots, and this cleans the water (Van der Merwe in Krige 2018). However, if not managed, reeds can halt the flow of water in certain parts of the river, causing excessive flooding.

Fig. 3.14 The Blesbokspruit wetland. (Photograph by Stan Madden)

Based on these human-induced changes, the Blesbokspruit is described as an artificial wetland that was created through mining (Blesbokspruit Trust representatives, personal communication, 2018; Govender, personal interview, 2018; Madden, personal interview, 2018). It developed into a permanent wetland because of the discharge of large quantities of water from underground gold mines in the early 1990s (Ambani 2013, p. 19). This means that the quantity of water in the Blesbokspruit is no longer dependent on the season but has water all-year-round. The Blesbokspruit has therefore been partially modified due to the impacts on the river and serves as a buffer for the water entering the Vaal River, the main source of water for Gauteng's socio-economic activities (Ambani and Annegarn 2015, p. 48). The Blesbokspruit wetland acts as a purifier of mining, industrial and wastewater treatment works discharges. The challenge is now for water management, especially in agriculture due to the importance of the sector for food security. Much stricter inspections by decision-makers and civil society of new infrastructure development in relatively open river basins are needed to avoid over-commitment of already limited water resources (Molle et al. 2007, p. 585). Water resources in the Blesbokspruit are fully committed to a variety of human uses. This implies that water quality is degraded, river-dependent ecosystems are endangered and increasing demand for water is leading to strong competition between users (Molle et al. 2007, p. 587).

Owing to the over-exertion on the system and its deteriorated state, in 1996 the Blesbokspruit wetland was placed on the Montreux Record. This is equivalent to a person being blacklisted for having unpaid debt (Ambani and Annegarn 2015, p. 1; *Spring Advertiser* 2014b). Environmental sites of international importance that experience serious ecological problems both for humans and wildlife are placed on this record. It is "a register of wetland sites on the List of Wetlands of International Importance where changes in ecological conditions have occurred, are occurring, or are likely to occur as a result of technological developments, pollution or other human interference" (Ramsar website 2017). Despite numerous attempts to better the Blesbokspruit wetland, it remained on the Montreux Record throughout the duration of the research conducted for this book. Mashau (in Blesbokspruit Forum 2018d), a representative of the Gauteng Department of Agriculture and Rural Development (GDARD), explained that in order for the Blesbokspruit to be removed from the Montreux Record, it has to regain its Ramsar status and meet specific water quality guidelines. Mashau (in Blesbokspruit Forum 2019a, d) further reported that the paperwork to remove the Blesbokspruit from the record was complete and had been submitted to the Department of Environmental Affairs (DEA), who is responsible for submitting the paperwork to the National Ramsar Convention on behalf of GDARD. This was a positive attempt from the GDARD. However, by the end of the research period the DEA had still not submitted the documents to Ramsar.

3.5 Non-statutory Bodies Contributing to the Governance of the Blesbokspruit

This section introduces two non-statutory bodies in the management of the Blesbokspruit catchment. It is important to note that this section does not explain the direct roles of the various departments that are involved in managing the Blesbokspruit; this is explained in Chap. 7 (Sect. 7.3.1) to provide a detailed background on the roles and responsibilities of the various spheres of government with regard to the management of the Blesbokspruit, before delving into the issues of power relations between governing bodies and other stakeholders (which is the purpose of Chap. 7). This section specifically refers to the role of the Blesbokspruit Forum, which is attended by government departments that manage the Blesbokspruit catchment. They meet with interested parties, sectors and businesses that interact with the Blesbokspruit through their discharge of water or the abstraction of water. This forum is significant to the book because all stakeholders who interact with the Blesbokspruit are meant to attend this forum to report back to create an awareness of what takes place in the catchment, therefore, to enhance communication pertaining to the Blesbokspruit catchment. It is a public forum and is open to the community. The Blesbokspruit Trust is a non-governmental organisation (NGO); their contributing role to the catchment is explained below.

Members of the Blesbokspruit Wetland Forum, the Gauteng Wetland Forum and Blesbokspruit Trust served as significant sources of information for the research. The Blesbokspruit Forum (discussed further in Sect. 3.5.1) is a catchment forum where representatives of government departments present their water quality results according to a set of In-Stream Water Quality Guidelines, and organisations have to report on their water usage (Reservoir website 2020; Blesbokspruit Forum Charter

2003). A Department of Water and Sanitation (DWS) representative records the information and distributes the minutes of the meetings to the forum representatives. Meetings take place in Springs at the Grootvaly Educational Centre (Blesbokspruit Forum 2018a, b, c, d)[5]. The Gauteng Wetland Forum is a platform where relevant representatives of the various catchment forums in Gauteng meet to provide feedback on each catchment. The vision of the Gauteng Wetland Forum is to "effectively conserve and manage wetland ecosystems in the Gauteng Province" (Wetlands Portal of South Africa 2016). Meetings take place at the Endangered Wildlife Trust venue in Modderfontein (Gauteng Wetland Forum 2018a, b, c). These meetings allowed for sound interaction with various individuals among the stakeholders (Fig. 3.15). The Blesbokspruit Trust (discussed further in Sect. 3.5.2) consists of several members who are residents in the area. They have a vested interest in protecting the Blesbokspruit Ramsar site because they reside in the area. The Trust consists of 10–12 members (five of whom were interviewed) who meet on a monthly basis to discuss the welfare of the area and address issues pertaining to the Blesbokspruit. For instance, the excessive reeds and hyacinth growth halt the flow

Fig. 3.15 Members of the Gauteng Wetland Forum during a tour of Marievale Bird Sanctuary led by Stan Madden. (Photographs by researcher)

[5] During the Covid-19 global pandemic, all forum meetings took place online via Zoom.

of the Blesbokspruit; therefore, maintenance of the reeds needs to take place (Naidoo, personal interview, 2018).

3.5.1 *The Blesbokspruit Forum*

The following information is taken from the Blesbokspruit Forum Charter (2003, p. 2). The Blesbokspruit Forum was established in 1996, and since the adoption of the National Water Act (NWA) in 1998, its intention was to promote the aims of the NWA. The Forum's mission is "to provide a platform to assist in the development of an integrated environmental management strategy for the Blesbokspruit catchment through stakeholder participation". The Blesbokspruit Forum is a body that meets on a quarterly basis, providing an opportunity for all role-players to participate on a platform that is transparent to discuss the issues within the Blesbokspruit catchment. Therefore, "the participation of all people in the protection, use, development, conservation, management and control of the water resources of the Blesbokspruit catchment" is promoted through this forum (Blesbokspruit Forum Charter 2003, p. 2–3). It serves as the biggest communication link between the spheres of government related to the Blesbokspruit (Maurizi, personal interview, 2018). Water use licences, environmental authorisations and mining rights are presented and discussed in the forum. The aim is for all information pertaining to the management of the catchment to be made available to stakeholders. It is therefore a body that "has the capacity to make recommendations to the authorities and other forum management structures [such as the Gauteng Wetland Forum] on behalf of the broader body of Forum members". However, this Forum, according to its charter, cannot be considered a pressure group, an activist body or one that can dictate actions to participants (Blesbokspruit Forum Charter 2003, p. 2–3). The Forum consists of government departments concerned (i.e. national, local and provincial), mines, industries, farmers, local authorities, NGOs, water service providers and the general public. Considering that this forum brings together stakeholders that contribute and have a role in managing the Blesbokspruit catchment, this is a platform where communication is transferred; therefore, the information about the Blesbokspruit and the water quality is presented on this platform. If a stakeholder or public individual is not aware of why a water use licence, for instance, was issued, they could question the relevant authority at a forum meeting or through the e-mailing group.

Through observation and attendance of forum meetings since 2018, it was found that the Blesbokspruit Forum is chaired[6] by the DWS and is made up of representatives of 18 organisations, including mining companies, industry and the agricultural sector, and government units. Local government includes the CoE and Lesedi Local Municipality, provincial government includes GDARD and national government

[6] Rand Water assisted in chairing forum meetings since 2020, once they were held virtually.

includes the Department of Mineral Resources (DMR) and the DWS. Five units within or under the DWS are represented, which are the Trans-Caledon Tunnel Authority (TCTA), Compliance Monitoring and Enforcement (CME), Catchment Management Agencies, Water Services Regulation and Water Use Licences. Lastly, two NGO representatives of the Gauteng Wetland Forum and the Blesbokspruit Trust are also typically present (Blesbokspruit Forum 2018a, b, c, d, 2019a, b, c, d). Six water quality reports are normally presented at forum meetings by the DWS, the CoE, the Ekurhuleni Water Care Company (ERWAT), Lesedi Local Municipality, Rand Water[7] and the TCTA. The forum is open to public attendance, and influential members of the community attend to raise issues on behalf of the community. From a government point of view, this is considered part of the public participation process for environmental conservation. As of 2020 points were allocated to those who participate in the forum, by the South Africa National Accreditation System, each representative is able to obtain two points a year (half a point allocated for attending the four meetings per year). This is significant to those who are involved in managing the Blesbokspruit to indicate their attendance and contribution to the forum. Further, those representatives who do not usually attend forum meetings might be more inclined to attend as a result.

Numerous presentations typically take place in the forum, such as new environmental authorisation processes, new mining applications, water use licence status in the catchment, environmental conservation and wetland protection. Further, pollution incidents are a permanent agenda item for raising current or new incidents. Details of the Blesbokspruit Forum, such as the minutes and water quality reports, are placed on the Reservoir website, which is administered by Rand Water on behalf of the DWS (Blesbokspruit Forum 2018a). At the beginning of 2019, a Google e-mailing group was created to ensure that the transfer of information outside forum meetings reached all stakeholders involved in the Blesbokspruit catchment. This platform allows stakeholders to raise issues before and after meetings (Blesbokspruit Forum 2018d).

This forum was therefore crucial to the research because it served as the most important platform where communication between stakeholders took place. Those who attended forum meetings were also able to share the information with other stakeholders and therefore contributed significantly as a factor (the transfer of information and communication) that influenced key individuals' and stakeholders' social constructions of the water quality of the Blesbokspruit (discussed in Chap. 6).

3.5.2 *The Blesbokspruit Trust*

Members who live in Springs (see Fig. 3.16) and who were interested in environmental conservation formed the Trust as a registered NGO in 1998 (Madden, personal interview, 2018). At the time that this research was conducted, Charl van der Merwe served as the chairperson of the Trust (Blesbokspruit Forum 2019a). The

[7] Rand Water is the bulk water provider in the Gauteng Province and surrounding provinces.

Fig. 3.16 Blesbokspruit Trust members (from left to right: Ewald Meyer, Charl van der Merwe and Stan Madden). (Photograph by researcher)

main aim of the Trust is for its voluntary members to contribute to reducing pollution impacts on the Blesbokspruit by reporting these incidents to the appropriate government department. Often, residential property owners contact the Trust regarding pollution incidents and they contact the municipality to ensure that the problem is attended to (Van der Merwe, personal interview, 2018). Members of the Trust live in residential areas adjacent to the Blesbokspruit, and they have a vested interest in issues pertaining to the river. The Trust assists with reviewing environmental impact assessments (EIA) and attending stakeholder meetings regarding new mining applications, as well as being part of the Blesbokspruit Forum and Gauteng Wetland Forum.

Members of the Trust meet monthly, and they function similarly to an NGO. Initially, the Trust was made up of residential property owners some of whom were also representatives of corporate companies, such as the SAPPI Enstra Mill, Impala Platinum, Grootvlei Mine and Zincor Operation, and government officials. Residential property owners who are members of the Trust got their respective companies involved because this contributed to their livelihood and benefitted these companies by conserving the environment. Some of these corporate companies sponsored the Trust (such as Zincor) via monthly financial contributions over a period of 10 years (Van der Merwe, personal interview, 2018). The Trust relies on sponsors and donations to operate, which has declined over the years (Naidoo, personal interview, 2018). The Trust was financially sound when it began but is currently struggling to maintain its sponsors. Anglo American is one of the companies that no longer sponsors the Trust, and by 2018 there were severe financial constraints (Naidoo, personal interview, 2018). While Carnival City[8] is the biggest funder and Impala Platinum also continues to fund the trust (Naidoo, personal interview, 2018), it is a struggle to secure sufficient finances. Many members are leaving the Trust due to the closure of mines, which places financial strain on companies. The corporate companies that still serve as members are not in a financial position to provide ongoing financial assistance (De Jager, personal interview, 2018). According to Pravin Naidoo, who is a member of the Trust and a Springs resident and manager at Impala Platinum, Impala is an ongoing sponsor of the Trust. All the sponsored money goes towards environmental conservation.

An environmental education centre, called the *Grootvaly Educational Centre*, was built through the CoE, with the assistance of many sponsors, such as Impala Platinum[9]. The Trust runs this centre, built to train children, especially the underprivileged, in environmental conservation (Naidoo, personal interview, 2018). Busloads of schoolchildren were taught at the centre once a week, through funds made available by Anglo American (Madden, personal interview, 2018). In 2017, it educated 1411 children, but owing to the high transport costs involved in bringing children to the centre, the number of children has reduced over the years (Van der Merwe, personal interview, 2018). The Trust also consistently tries to make many people aware of what they do (De Jager, personal interview, 2018; Naidoo, personal interview, 2018).

3.6 Research Methodology

The Blesbokspruit was used as the case study for the research conducted for this book. A case study was used to explore the views of various key individuals and other stakeholders who interacted with the Blesbokspruit due to their need for water and based on their need for water, how they socially constructed the quality of the water. A case study was the most suitable approach for this research to identify stakeholders' perceptions of AMD treatment in the eastern basin, and how and why various social constructions of the water quality of the Blesbokspruit were formed due to vested interests, which resulted from power relations.

Owing to the in-depth nature of the research, a qualitative research design was used to obtain both primary and secondary data. The book therefore uses a qualitative research design to explain how water quality in the Blesbokspruit is socially constructed in the context of AMD and its treatment.

---

[8] Carnival City is a hotel and casino situated in the East Rand.

[9] According to Naidoo (personal interview, 2018), Impala is passionate about the environment and spent close to a billion rand over the last 20 years to assist with environmental conservation in the area.

3.6.1 Data Collection

The data collection included a combination of primary and secondary data. Data were collected in two phases during the research which stretched from January 2018 to August 2020. The first phase involved identifying the various stakeholders who interact with the Blesbokspruit and the key individuals within stakeholders who were linked to AMD treatment. The second phase involved conducting in-depth interviews, attending the Blesbokspruit Forum and the Gauteng Wetland Forum, and site visits.

Primary data included scoping interviews, face-to-face in-depth interviews with key individuals, attendance of Blesbokspruit Forum meetings (since 2018), attendance of Gauteng Wetland Forum meetings (since 2018), attendance of the Blesbokspruit Trust meeting, attendance of the 2017 launch of the eastern basin AMD treatment plant. Many site visits to the East Rand were conducted, which included the Blesbokspruit wetland, Springs and Nigel communities, mining areas and farms, a tour of the AMD treatment plant, visits to the Stable Inn Conference Venue and the Riverside Conference and Wedding Venue, and the Marievale Bird Sanctuary with the Gauteng Wetland Forum. Personal and follow-up communication with key individuals and other stakeholders took place via telephone and e-mail, before or after forum meetings and after site visits. Telephone communications with Springs and Nigel community members (who formed part of the EIA process in 2014) were conducted on their post-views of the AMD STT. Observations of key individuals' interactions with the Blesbokspruit took place during site visits and observations of stakeholders during the forum meetings.

The purpose of conducting follow-up communication with community members who formed part of the EIA process was to identify whether they held the same views after the STT treatment began, as they did during the EIA process[10]. Some of the key individuals that were interviewed served as key informants, providing information on public meetings, new mining activities or communication between government officials regarding pollution incidents and providing links to other participants for the research. All participants used for the research were adults over the age of 18 years. Attendance of the Blesbokspruit Forum and Gauteng Wetland Forum, and site visits to the AMD treatment plant, the Blesbokspruit, agricultural areas, surrounding residential areas and tourist attractions formed a significant part of the research.

Secondary data included a combination of published literature (books, journal articles) on the Blesbokspruit, media reports, official government documents (including the EIA for the sludge disposals site and TCTA reports), national legislation, acts and policies pertaining to water, mining and the environment. Key

[10]The prominent community members interviewed for this study are part of the key interviews conducted because they are influential in the community; they attend the catchment forums, form part of the Blesbokspruit Trust and often referred to in media reports. The community members who were telephonically interviewed were chosen based on the comments they provided during the EIA process because they are property owners.

interviewees provided published and supposed publicly available documents that were not easily accessible, which assisted with the research. Such documents were useful in cases where some people did not want to be interviewed because they did not have permission from their department heads, for example.

In 2016, five scoping interviews were conducted with stakeholders. This included an environmental activist, Ms Mariette Liefferink, Chief Executive Officer (CEO) of the Federation for a Sustainable Environment, and four stakeholders within the agricultural sector, namely, Mr Bennie van Zyl from the Transvaal Agricultural Union, a farmer on the West Rand (at this stage the research area was still to be confirmed), Dr Piet Nell from the Agricultural Research Council (ARC) and an agricultural consultant on the East Rand. From these scoping interviews and reading newspaper articles, it was clear that the major topic of discussion with regard to the Blesbokspruit was on AMD treatment and its potential implications (or not) for the Blesbokspruit. The AMD STT in the eastern basin was launched in February 2017, and exploring and understanding the perceptions regarding the water quality of the Blesbokspruit in the context of the AMD STT proved relevant to various stakeholders. The scoping interviews assisted in formulating the research objectives. In addition, extensively reviewing the EIA documents compiled by Digby Wells Environmental early in the research process was essential to form a better understanding of the views of the various stakeholders on the proposed AMD STT and its anticipated impacts on the quality of water of the Blesbokspruit. Numerous stakeholders felt that their views were not being heard and that the EIA process was merely a tickbox exercise. The EIA documents indicated who the key individuals were among stakeholders and which areas were likely to be most affected by the establishment of the AMD treatment plant.

The section that follows provides a brief explanation of how the participants were selected, how the in-depth interviews were conducted and how observations took place. It is important to note that based on the variety of data collection methods used in this book the findings of the book represent the view of key individuals and stakeholders in general that are linked to the Blesbokspruit.

3.6.2 Selection of Participants, In-Depth Interviews and Observations

Purposive sampling and snowball sampling were used in this research to identify key individuals. *Key* is explained as those individuals among stakeholders who have expert knowledge on water quality and AMD treatment; those involved in the management of the Blesbokspruit catchment; those using the water of the Blesbokspruit; and those who live in the surrounding areas. This book does not claim that the individuals interviewed are a representative sample of a particular stakeholder group (e.g. tourism, agriculture, industry, community) or sphere of government or particular government department. The intention was not to identify a representative sample, but rather to conduct interviews with key individuals. Even though key individuals were interviewed, it is not entirely an individualised process. It is possible that these individuals' thinking about the Blesbokspruit, and the treatment of AMD, is linked to, and influenced by, their stakeholder grouping.

Through reviewing media reports and official government documents and policies, further relevant individuals were identified. The scoping interviews also assisted in identifying additional individuals who were beneficial sources of information. Attendance of the catchment forums made it possible to identify other appropriate individuals for the research, that is, those who attended the meetings, what role they played and whether they would be beneficial to the research.

Even though purposive sampling was used, part of the sample was determined as the research progressed. For one, snowball sampling was used to find further relevant individuals. Snowballing worked well where an interview was requested with a specific individual, but the person was hesitant due to his or her organisational affiliation. In such cases, they provided useful suggestions to alternative people. After becoming familiar with who was directly involved in the AMD STT project team and who key individuals in the Springs community were, they too were approached for interviews. Other helpful community members were identified through the EIA documents reviewed for the research. Those members who frequently commented during the EIA public participation process and made comments that were valid for this research were contacted.

Each interview was structured based on the individuals' professional stance and affiliation to the Blesbokspruit catchment, and what role they played with regard to AMD STT and the management of the Blesbokspruit. It is important to note that some of the interviewees play more than one role in the Blesbokspruit due to various interests; for instance, they may live and work in the area. Some key individuals are associated with more than one stakeholder; for instance, those who are both residents and working for a stakeholder, such as the City of Ekurhuleni [CoE; formerly Ekurhuleni Metropolitan Municipality (EMM)]. Others owned property and had businesses in the tourism sector. Most of the prominent community members interviewed also played an environmental activist role. The information collected from each individual was based on these roles, and each response was used based on these roles to identify how social constructions are formed.

Observations were made during the catchment forums; at the launch of the AMD treatment plant; during site visits to the Blesbokspruit and surrounding mining and agricultural land, tourist attractions, including the Marievale Bird Sanctuary, and residential areas; and during a tour of the AMD treatment plant.

3.7 Conclusion

This chapter provided a background on the East Rand, the eastern basin, and the Blesbokspruit and its wetland. The relevance of this chapter was to place the Blesbokspruit in the context of mining on the East Rand and to showcase the importance of the Blesbokspruit based on the various uses of the water. The known water uses can be grouped into five categories, namely (1) agriculture, (2) mining, (3) industry, (4) tourism and (5) domestic use. The next chapter explains how AMD became an issue in South Africa and, more specifically, the eastern basin, requiring AMD treatment to reduce its impact on the Blesbokspruit and its users of the water.

References

Adler, R., Claasen, M., Godfrey, L. & Turton, A. 2007. Water, mining and waste: An historical and economic perspective on conflict management in South Africa. *The Economics of Peace and Security Journal*, 2(2):33–41.

Ambani, AE. 2013. *Long-term assessment of the surface water quality in the Blesbokspruit Ramsar wetland*. Unpublished MSc dissertation. Johannesburg: University of Johannesburg. Available at: https://ujcontent.uj.ac.za/vital/access/manager/Repository/uj:11619. Accessed on: 21 June 2019.

Ambani, AE. & Annesgarn, H. 2015. A reduction in mining and industrial effluents in the Blesbokspruit Ramsar wetland, South Africa: Has the quality of the surface water in the wetland improved? *Water SA*, 41(5):648–659. Available at: https://journals.co.za/content/waters/41/5/EJC179283. Accessed on: 22 February 2017.

Blesbokspruit Forum. 2018a. Minutes of Blesbokspruit Forum meeting and personal attendance, 8 February 2018.

Blesbokspruit Forum. 2018b. Minutes of Blesbokspruit Forum meeting and personal attendance, 11 May 2018.

Blesbokspruit Forum. 2018c. Minutes of Blesbokspruit forum meeting and personal attendance, 8 August 2018.

Blesbokspruit Forum. 2018d. Minutes of Blesbokspruit Forum meeting and personal attendance, 8 November 2018.

Blesbokspruit Forum. 2019a. Minutes of Blesbokspruit Forum meeting and personal attendance, 7 February 2019.

Blesbokspruit Forum. 2019b. Minutes of Blesbokspruit Forum meeting, 9 May 2019.

Blesbokspruit Forum. 2019c. Minutes of Blesbokspruit Forum meeting, 8 August 2019.

Blesbokspruit Forum. 2019d. Minutes of Blesbokspruit Forum meeting and personal attendance, 7 November 2019.

Blesbokspruit Forum Charter. 2003. Terms of reference for the Blesbokspruit Forum October 2003. Available at: http://www.reservoir.co.za/forums/vaalbarrage/blesbok_forum/blesbok_documents/BF_TOR_Oct2003.pdf. Accessed on: 1 July 2018.

Blesbokspruit Trust Representatives (Meyer, E., Madden, S. & Van der Merwe, C). 2018. Blesbokspruit Trustee Members. Personal communication, 8 February. Springs, South Africa.

Creamer, M. 2019a. Menar's Canyon offers Eskom cheaper coal on open-book basis. *Mining Weekly,* 19 March. Available at: http://www.miningweekly.com/article/menars-canyon-offers-eskom-cheaper-coal-on-open-book-basis-2019-03-19. Accessed on: 4 April 2019.

Creamer, M. 2019b. Researchers at Mandela Mining Precinct deserve strong support. *Mining Weekly,* 29 May. Available at: https://www.miningweekly.com/article/research-development-and-innovation-activists-at-mandela-mining-precinct-deserve-strong-support-2019-05-29/rep_id:3650 Accessed on: 30 May 2019.

De Jager, P. 2018. Lawyer and Springs resident. Personal interview, 25 January. Springs, South Africa.

Digby Wells Environmental. 2015. *Construction and operation of the proposed sludge disposal facility and pipelines associated with the treatment of acid mine drainage in the eastern basin of the Witwatersrand, Gauteng: Final environmental impact assessment report.* Johannesburg: Digby Wells and Associates.

du Plessis. A., Harmse, T. & Ahmed, F. 2014. Quantifying and predicting the water quality associated with land cover change: A case study of the Blesbok Spruit catchment. *South Africa Water*, 6(10):2976–2968. Available at: https://www.mdpi.com/2073-4441/6/10/2946/htm. Accessed on: 14 August 2020.

Ekurhuleni Metropolitan Municipality (EMM). 2007a. *Environmental management framework for Ekurhuleni*. Ekurhuleni: EMM. Available at: https://www.ekurhuleni.gov.za/yourservices/environmental-health/environmental-policies-plans-and-strategies/469-environmental-managment-framework-2008/file.html. Accessed on: 15 October 2018.

Ekurhuleni Metropolitan Municipality (EMM). 2007b. *Wetland inventory report: Identification, classification, assessment & delineation of wetlands within the Ekurhuleni Metropolitan Municipality*. Ekurhuleni: EMM. Available at: https://www.ekurhuleni.gov.za/473-env-wetland-inventory-report-2007/file. Accessed on: 15 October 2018.

Environomics. 2014. *Gauteng environmental management framework report*. Johannesburg: Gauteng Province, Agriculture and Rural Development. Available at: http://www.gauteng.gov.za/government/departments/agriculture-and-rural-development/Documents/Gauteng%20Provincial%20Environmental%20Management%20Framework/2014%20GPEMF%20Section%20A%20Introduction%20and%20Status%20Quo.pdf. Accessed on: 20 June 2019.

Gauteng Tourism Authority. 2020. Blesbokspruit Gauteng Tourism Authority. Available at: https://www.gauteng.net/attractions/blesbokspruit. Accessed on: 5 March 2020.

Gauteng Wetland Forum. 2018a. Minutes of Gauteng Wetland Forum meeting and personal attendance, 23 February 2018.

Gauteng Wetland Forum. 2018b. Minutes of Gauteng Wetland Forum meeting and personal attendance, 25 May 2018.

Gauteng Wetland Forum. 2018c. Minutes of Gauteng Wetland Forum meeting and personal attendance, 31 August 2018.

Google Maps. 2020. Map of the stakeholders who interact with the Blesbokspruit. Available at: https://www.google.com/maps/dir/Acid+Mine+Drainage+Plant+-+Eastern+Basin. Accessed on: 30 September 2019.

Govender, B. 2018. Chief directorate Mine water management unit: Department of Water and Sanitation. Personal interview, 1 March. Pretoria, South Africa.

Hartnady, CJH. 2009. South Africa's gold production and reserves. *South African Journal of Science*, 105. Available at: archive.sajs.co.za/index.php/SAJS/article/view/101/82. Accessed on: 11 June 2019.

Hawley, G. & Desmet, P. 2020. *Review of the City of Ekurhuleni Bioregional Plan 2020*. Draft for public comment, July 2020.

Kirk, J., Ruiz, J., Chelsey, J. & Titley, S. 2004. The origin of gold in South Africa. *American Scientist*, 91. Available at: https://www.researchgate.net/publication/240968643. Accessed on: 11 June 2019.

Krige, I. 2018. Walk and talk at the Blesbokspruit. *Springs Advertiser,* 9 July. Available at: https://springsadvertiser.co.za/192898/walk-and-talk-at-the-blesbokspruit/. Accessed on: 10 August 2018.

Labuschagne, DC. 2015. *A 3D geological model for the East Rand Basin, South Africa*. MSc dissertation. Potchefstroom: North West University. Available at: https://dspace.nwu.ac.za/bitstream/handle/10394/15351/Labuschagne_DC.pdf?sequence=1&isAllowed=y. Accessed on: 4 October 2018.

Madden, S. 2018. Environmentalist and Springs resident. Personal interview, 8 February. Springs, South Africa.

Milaras, M., Ahmed, F. & McKay, TJM. 2014. *Mine closure in South Africa: A survey of current profession thinking and practice*. Johannesburg: University of the Witwatersrand. Available at: https://www.researchgate.net/publication/278035528. Accessed on: 7 May 2020.

Milwaukee Riverkeeper. 2015. What's a river basin? What's a watershed? Available at: https://www.milwaukeeriverkeeper.org/whats-a-river-basin-whats-a-watershed/. Accessed on: 5 September 2019.

Molle, F., Wester, P, Hirsch, P., Jensen, JR., Murray-Rust, H., Paranjpye, V., Pollard, S & Van der Zaag, P. 2007. River basin development and management. In *Water for life: A comprehensive assessment of water management in agriculture.* Edited by Molden, D. London; Earthscan: 585-624. Available at: https://www.researchgate.net/publication/40794254_River_basin_development_and_management. Accessed on: 5 September 2019.

Naidoo, P. 2018. Manager Platinum and Metals: Impala Platinum. Personal interview, 7 March. Springs, South Africa.

Neingo, PN. & Tholana, T. 2016. Trends in productivity in the South African gold mining industry. *The Journal of the Southern African Institute of Mining and Metallurgy*, 116. Available at: http://www.scielo.org.za/pdf/jsaimm/v116n3/14.pdf. Accessed on: 11 June 2019.

Pillay, M. 2018. Business Development Executive: Digby Wells Environmental. Personal interview, 16 January. Bryanston, South Africa.

Ramsar. 2017. *The Ramsar Convention on Wetlands: The Montreux Record.* Available at: http://archive.ramsar.org/cda/en/ramsar-documents-montreux/main/ramsar/1-31-118_4000_0__. Accessed on: 10 January 2018.

Republic of South Africa (RSA). 1998. *National Water Act, no 36 of 1998.* Pretoria: Government Printer. Available at: http://www.dwaf.gov.za/Documents/Legislature/nw_act/NWA.pdf. Accessed on: 14 April 2013.

Reservoir website. 2020. About the Reservoir website. Available at: http://www.reservoir.co.za/about.htm. Accessed on: 13 August 2020.

Scott, R. 1995. *Flooding of the Central and East Rand gold mines: An investigation into controls over the inflow rate, water quality and predicted impacts of flooded mines* (Water Research Report No. 486/1/95). Pretoria: Water Research Commission. Available at: http://www.wrc.org.za/wp-content/uploads/mdocs/486-1-95.pdf. Accessed on: 11 June 2019.

South African Cities Network. 2019. City of Ekurhuleni. Available at: http://www.sacities.net/member-cities/ekurhuleni. Accessed on: 4 September 2019.

South Africa Venues. 2019. Grootvaly Wetland Reserve. Available at: https://www.sa-venues.com/game-reserves/grootvaly-wetland-reserve.php. Accessed on 4 September 2019.

Springs Advertiser. 2014a. Blesbokspruit gets too much water: The nutrient-rich water caused the reeds to overgrown the spruit. *Springs Advertiser*, 13 April. Available at: http://springsadvertiser.co.za/89328/blesbokspruit-gets-too-much-water/. Accessed on: 22 February 2017.

Springs Advertiser. 2014b. Springs says no to opencast mining: Residents oppose these-zoning of the land. *Springs Advertiser*, 23 April. Available at: https://springsadvertiser.co.za/90116/springs-says-opencast-mining/. Accessed on: 22 February 2017.

Storey, A. 2018. Nigel resident. Personal interview and site visit, 26 January. Nigel, South Africa.

Tempelhoff, J., Munnik, V. & Viljoen, M. 2007. The Vaal Barrage, South Africa's hardest working water way: An historical contemplation. *The Journal for Transdisciplinary Research in Southern Africa*, 3(1):107–133.

Thorius, T. 2004. *The effect of Grootvlei mine water on the Blesbokspruit.* MSc dissertation on Environmental Management. Johannesburg: University of Johannesburg. Available at: https://ujcontent.uj.ac.za/vital/access/manager/Repository/uj:14868. Accessed on: 8 July 2019.

Tucker, RF., Viljoen, R. & Viljoen, MJ. 2016. A review of the Witwatersrand Basin: The world's greatest goldfield. *Episodes*, 39(2):105–133. Available at: https://www.researchgate.net/publication/305924249_A_Review_of_the_Witwatersrand_Basin_-_The_Worlds_Greatest_Goldfield. Accessed on: 11 June 2019.

Van der Merwe, C. 2018. Chairman: Blesbokspruit Trust. Personal interview, 21 February. Springs, South Africa.

Chapter 4
Acid Mine Drainage and Its Treatment in the Eastern Basin

Abstract This chapter looks at acid mine drainage (AMD) in South Africa and, specifically, the history of mining in the eastern basin of the Witwatersrand, which is the focus area. This chapter aims to contextualise the problems that mining led to in the form of AMD. The main purpose of this chapter is to explain the short-term treatment (STT) of AMD and how it was implemented, the launch of the eastern basin high-density sludge treatment plant and to present the environmental impact assessment (EIA) process for the STT. This chapter provides a background on how the treatment was planned and implemented, which is necessary to understand the perceptions of the AMD STT.

Keywords Acid mine drainage treatment plant · Environmental impact assessment · Sludge disposal

4.1 Introduction

Mining is an essential component for economic development, and water is an essential component for sustainable development. How do we balance these two equally crucial resources to achieve the aim of sustainable development, which is to enhance both environmental security and human security? Gold mining on the East Rand created an abundance of economic wealth, and now coal – used for electricity generation – and its mining is said to contribute to economic growth. Yet, current and future open-cast coal mining activities create an uproar among communities and environmental activists, causing these stakeholders to revolt at new mining applications. This is due to the fact that extensive environmental damage has occurred on the East Rand and impacted those who have lived in the area for many years. These negative impacts have led to a stark awareness among the community, considering the eastern basin is undergoing short-term acid mine drainage (AMD) treatment to rectify the damages caused by past mining activities.

S. Naidoo, *Social Constructions of Water Quality in South Africa*,
https://doi.org/10.1007/978-3-030-98237-9_4

This chapter looks at AMD in South Africa and, specifically, the history of mining in the eastern basin of the Witwatersrand, which is the focus area of the book. This chapter aims to contextualise the problems that mining led to in the form of AMD. The main purpose of this chapter is to explain the short-term treatment (STT) of AMD and how it was implemented, the launch of the eastern basin high-density sludge treatment plant and to present the environmental impact assessment (EIA) process for the STT. The chapter will provide a background to how the treatment was planned and implemented, which is necessary to understand the perceptions of the AMD STT (discussed in Chap. 5).

4.2 History of Acid Mine Drainage in South Africa and Why It Occurs

Where there is mining, there is the risk of AMD. AMD is directly caused and linked to mine sites that are no longer in operation, either due to the end of a mining operation or because the mining company was liquidated (Adler et al. 2007; Coetzee et al. 2010; Guedes 2010; Liefferink 2012; Naidoo 2017). Vast gold resources were discovered in South Africa in 1886 on the Witwatersrand, allowing the mining industry to play a central role in the country's economic, social and political environment (Adler et al. 2007, p. 34). Since then, Johannesburg has progressed from a "dusty mining town to a major urban and industrial conurbation" that provided housing to a quarter of the population of South Africa (Turton et al. 2006, p. 313). According to Turton and colleagues, this accounted for a contributory percentage of economic activity on the African continent, and therefore, sustaining the population and South Africa's minerals are highly diversified, profitable and plentiful, which led to the government allowing the mining industry to maximise its profits and externalise costs. "As a city founded on mining, Johannesburg's legacy is the wealth created, as well as the damage left behind by the careless mining practices of the past" (Guedes 2010, p. 67).

As mining declined over the years, there was a reduction in pumping and water filling up in the mine voids, which transferred to neighbouring mines due to the known interconnection of the mines on the Witwatersrand. "The simple reality is what came out of the void contains certain metals and elements that were laid down in geological times that were typically anoxic. This manifests as pyrite in the ore body and pyrite that oxidises and drives the chemical reaction known eventually as AMD"[1] (Turton, personal communication, 2018). The reduction in pumping meant that mines that remained operational had to take on larger pumping responsibilities to continue their activities. Owing to a reduction in the number of mines that were

[1] By depositing barren tailings, or sludge containing species of calcium, this will raise the pH in the void. This is desirable. It will also remove oxygen from the void, creating anoxic conditions that prevent the proliferation of AMD.

pumping, the state assisted with pumping costs due to the large amount of water mainly from abandoned mines (Environomics 2014, p. 28; Scott 1995). The quality of the water that decants from the mine void is completely poor for the environment and human use.

Neglect of the environment has been immense even though the mining industry has significantly contributed to South Africa's economic growth. This environmental neglect caused by mining has created increased expenses, resulting in growing concerns. These growing concerns have led to increasingly stricter legislation being imposed on the mining industry to reduce its impact on the environment (DWS 2017). In the past, the public had to bear the cost, through taxes, for the rehabilitation of mining; however, there are now more stringent rules in place where mining companies are required to have a rehabilitation fund before they are given the go-ahead to continue with their operations, and this will be required for new mines before they are issued a mining licence (Naidoo 2017, p. 92). The Inter-Ministerial Committee (IMC) on Acid Mine Drainage has played a role in this regard; according to this committee (Coetzee et al. 2010, p. 37), mine flooding and, by implication, also AMD is not something new; this problem has been around for centuries. The reason why mine flooding is a concern now is due to the extent of the problem, and the fact that more mines are now left ownerless and abandoned than in the past. Once the mines are abandoned, pumping of the underground mine water stops and with the rising water levels, acidic water starts to form. Water specialist Anthony Turton (2015) believes that AMD should be understood in terms of the historical evolution of the government's policy towards mining. Turton stated that the crucial problem is that the existing mining policy reflected a historical legacy in which powerful financial interests coincided with the interests of a racially defined political elite. As a result, the government was viewed as becoming an agent rather than a controller of the mining industry (Naidoo 2017, p. 36). Therefore, AMD-related issues in South Africa are riddled with political struggles between stakeholders. Mpofu et al. (2018, p. 80) suggested that AMD issues could be a result of government and mining companies not acting in the best interests of the environment and, instead of working as partners, they were operating as separate entities working solely for their own interests, leading to increasing environmental impacts from mining in South Africa.

In South Africa there have been negative impacts on a number of areas, including the Witwatersrand goldfields, Mpumalanga Province (Carolina) and KwaZulu-Natal Province coal fields, and the O'Kiep copper district. However, Gauteng Province is said to be the most threatening as the western, eastern and central basins of the Vaal River system are regarded as the main areas of concern (Coetzee et al. 2010, p. 8). According to Pillay (personal interview, 2018), in the three basins there are about 200 million ℓ of acid water. Acidic water will devastate the quality of the water of the Vaal and Crocodile rivers, and, if not managed, it will affect all life dependent on those river systems (Parris 2010). Further, the various users that are abstracting water from the Vaal system, such as farmers and cattle, in turn increase the contamination levels (Pillay, personal interview, 2018). On the one hand, agriculture is a big polluter of water due to its dependence on a sustainable environment in the form of quality soil, water and atmosphere. On the other hand, this dependence is necessary to deliver products essential for human life (Parris 2010). Therefore, contaminated water can have a severe impact on both agriculture (producer of food) and on society (need for food) as a whole. Such consequences of AMD include impacts on people's health, plants and animals, and possible pollution of borehole water and soil used to grow vegetables (Environment 2016). This was evident in 2010, when the Grootvlei Mine was not treating and pumping the water (discussed in Sect. 4.3) according to their water use licence. Residents claimed that their gardens were fading; the bird life was affected; there was a decrease in the number of frogs, which are an indicator species given their vulnerability to a degraded environment; and a massive number of fish were floating on the banks of the river (*Mail & Guardian* 2010).

As a consequence, AMD from ownerless (liquidated and abandoned) mines created tremendous long-term environmental liabilities for the government (Hobbs et al. 2008, p. 417). The liability rests with the government because of its poor enforcement of legislation. The government, however, also remained adamant that it cannot be held responsible for the 'clean up' that is needed because of the high costs involved (Guedes 2010, p. 69). However, the government is footing the bill for the STT in the three basins of the Witwatersrand (Fig. 4.1), which is costing billions of rand. It is ultimately the South African public that is therefore partly covering the costs through increased water tariffs and taxes. The problem is that the STT involves only neutralising or partially treating[2] the water to increase the pH levels, but the sulphate levels remain high, which is a concern for river health (Mouton 2018). According to Bosman (personal interview, 2018), while she was a member of the Department of Water Affairs and Forestry (DWAF) [now Department of Water and

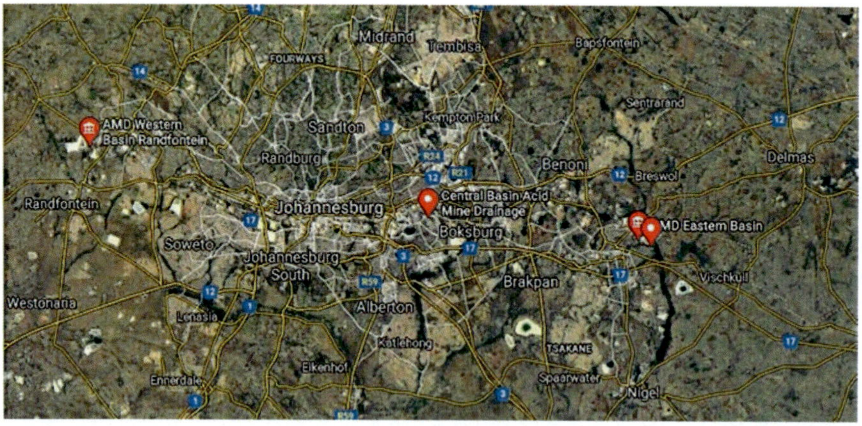

Fig. 4.1 Location of AMD treatment points in the Witwatersrand gold fields. (Google Maps 2020a)

[2]*Neutralisation* means the water is only partially or semi-treated, making it a short-term intervention.

Sanitation (DWS)], she suggested that AMD should have been treated to drinking standards shortly after it became a widely known issue in 2002. She was of the opinion that the problem could easily be addressed by removing the heavy metals, purifying the water and distributing the water to the people of Gauteng for drinking purposes. At this stage, she suggested that the DWS build a treatment plant because the impact of AMD could lead to serious possible contamination, such as from uranium and radioactivity.

In 1995 a publication by R. Scott on the treatment of discharged water indicated that "the discharged water whether pumped or allowed to emanate naturally will be of poor quality and some permanent treatment plant would be required to control the pH and remove dissolved solids from the water before discharge to a river system" (Scott 1995, p. xvii). Scott also indicated that the process of purifying the water demanded exorbitant sums of capital and required skilled labour for the operation to begin, and the process would have to operate indefinitely. Therefore, maintenance and upgrading systems could become a problem as well as the disposal of sludge and waste from the treatment process (Scott 1995, p. xvii). According to McKay and Milaras (2017, p. 1), in 1996, a treatment plan was presented to, and then accepted by, the Parliament in 1998, but little action was taken during this time until the STT began in the eastern basin in mid-2016. After mining ceased in the eastern basin in 2011, the South African government had to implement immediate short-term interventions to address the issue. The aim was to ensure that there would be no permanent damage (such as decanting of AMD) to the environment and to protect acidic water from reaching the Vaal system.

AMD is currently under control in the eastern basin through the STT interventions, which involves pumping of the water to keep it below the environmental critical level (ECL). However, because the STT involves only partially treating AMD water before its discharge into the natural environment, the quality of the water is still poor due to the high sulphate levels still prevalent in the discharged water. The long-term solutions have been set by the DWS, but not yet been implemented to address this environmental burden (Liefferink, personal interview, 2016). There are excessively high costs involved in treating AMD in the short term (Tlale, personal interview, 2018a). The increased expense is further evident by the fact that the implementation of the long-term treatment (LTT) is on hold due to lack of funding (Blesbokspruit Forum 2020) and due to "a recalibration of the salinity and hydrology model of the integrated Vaal River system" (Govender cited in Bega 2021). Therefore, the LTT is temporarily deferred and the STT will continue until funding for the LTT is secured. In the meantime, water experts stress the need to save what little water there is and to start the process of purifying contaminated water.

According to Govender (personal interview, 2018), in 2010 when the DWS started addressing AMD challenges, it realised that other areas being mined would also need mitigation strategies. These areas included Mpumalanga, Free State goldfields, North West platinum belt and some coal mining in Limpopo, including KwaZulu-Natal with abandoned coal fields, and if nothing was done, there would be very large-scale mine water impacts which would be bigger than what was experienced in Gauteng. The need to coordinate efforts is necessary in order to have a better understanding of what is going on across the country (Govender, personal interview, 2018). This led to the development of a draft Mine Water Management Policy (discussed further in Sect. 7.2). According to this draft policy (DWS 2017, p. 2), due to the long history of mining in South Africa there is still mineral wealth that is inaccessible across parts of South Africa. Considering the potential that this can have for local economic development and foreign investment, it was imperative that the DWS develop a policy principle to address mine water challenges. According to Craig Sheridan (cited in Mouton 2018), AMD is a massive problem, but if there is a strong political will behind addressing it, then it is not difficult to fix. Industrial wastewater contributes to water quality problems, but the two main sources that pollute South African rivers are AMD and poorly maintained wastewater treatment plants. The aim is to strengthen the protection of water resources from mine water contamination in the short- to long term (DWS 2017, p. 2). Further, the approach is to hold those parties accountable for negative effects and damage caused by AMD, to ensure South Africa is on track towards sustainable socio-economic development (*SAnews* 2017). The DWS completed a feasibility study to adopt the best plan of action for the LTT of AMD and found interim technologies proven for treatment of the expected volumes to the required standard. The idea is that the technology used must be situational based, sustainable, clean and economical (DWS 2017, p. 6). The next section focuses specifically on the occurrence of AMD in the eastern basin.

4.3 The Nature of AMD and Its Occurrence in the Eastern Basin

The mines on the East Rand are connected underground (Labuschagne 2015, p. 15; Madden, personal communication and site visit, 2018b; Scott 1995, p. 26), and underground water has historically been pumped from the eastern basin since the early mining activities in the 1940s and 1950s. As mines started closing from the 1950s and 1960s due to reaching their lifespan or liquidation, fewer contributions were made towards the pumping costs, which had to be undertaken by the few remaining mines. From 1963 onwards the state became more involved by granting subsidies or providing assistance to cover the pumping costs (Ambani 2013, p. 84; Scott 1995, p. 14). By 1991, only the SA Land and Exploration Gold Mining Company Mine and the Grootvlei Mine were in operation on the East Rand (Scott 1995, p. 4; Thorius 2004, p. 9; Van der Merwe and Lea 2003, p. 25), and the Grootvlei Mine took responsibility for the pumping costs (Ambani 2013, p. 84). The remainder of this section focuses on the Grootvlei Mine.

By 1995 the Grootvlei Mine could not manage the volume and cost of pumping and treating the water, and this led to large volumes of untreated AMD flowing into the Blesbokspruit and the mine's water use licence being withdrawn for non-compliance. This resulted in salination of soil and a decrease in the quality of the water (McKay and Milaras 2017, p. 1). In 1996 DWAF (now DWS) reissued a water use licence to the Grootvlei Mine, but after identifying that the mine was discharging untreated water into the Blesbokspruit, the licence was once again withdrawn. During 1996 the Grootvlei Mine was issued with a water use licence three times because the licence was valid for shorter periods in order to monitor and ensure that the conditions of the licence were met, and that the water was partially treated before being discharged into the Blesbokspruit (Ambani 2013, p. 85).

In 1997, despite the visible impacts on the water quality leading to the Blesbokspruit being placed on the Montreux Record the year before, due to the high sulphate levels evident in the water, a new water use licence was issued to the Grootvlei Mine, on condition that they establish a desalination (water purification) pilot plant. By 1998 most active mines on the Witwatersrand were contributing excessive amounts of saline underground water to an already impacted Vaal River system, implying that the Grootvlei Mine had to operate within their licence conditions to reduce the impact on the Vaal. In 1998 the DWS and Rand Water worked to develop a policy to limit the high sulphate levels (from discharged water) on the Vaal River, because it became evident that the Grootvlei Mine's operation was ultimately going to result in the discharge of AMD-contaminated water into the Vaal River. Water quality was poor due to artificial input from mines, wastewater treatment works and other industrial activities, caused by an increase in total dissolved salts (Ambani 2013, pp. 87–89).

In 1999, a fifth licence was issued to the Grootvlei Mine with the expectation that the mine would cover full desalination. However, by 2001, there were further issues, such as severe pump failures due to maintenance issues in the Grootvlei Mine's Shaft No. 3. The requisite high-density sludge plant to treat the underground mine water was not used. By 2002, the financial difficulties of Pamodzi Gold (owner of the Grootvlei Mine) were obvious due to the ongoing maintenance issues, and by 2009 several mines were under liquidation, including Pamodzi Gold (Ambani 2013, pp. 87–89). Pamdozi's Grootvlei Mine was sold in 2009 to Aurora Empowerment Systems, owned by Khulubuse Zuma (Chairperson) and Zondwa Mandela (Managing Director) (Crowley 2015; *Fin24* 2019; Holden, personal communication, 2019; Kings 2016; *Mail & Guardian* 2011). Aurora did not have sufficient funds to run the mine (Whittles 2016) and further lacked experience in mining management (Crowley 2015). Years prior to this there was intermittent pumping due to financial difficulties incurred by previous owners, and while Aurora operated the Grootvlei Mine, there was sporadic treatment (Holden, personal communication, 2019). While Aurora should have been pumping at 108 Mℓ/day the mine was only pumping at 40 Mℓ/day (*Mail & Guardian* 2010). In 2010, the mine was liquidated and ravaged by illegal miners, and more than 5000 employees lost their jobs and were left without pay (Crowley 2015; PMG 2010; Whittles 2016). Not one of the shafts at the Grootvlei Mine was left in working condition and shafts were either stripped or flooded (Evans 2014; *Mail & Guardian* 2011). In 2011, all that remained were the seven mine shafts (*Business Report* 2011).

In 2010, a *Mail & Guardian* article raised public awareness. Through overemphasis, the newspaper stressed the severity of the matter: "[T]he East Rand is an hour away from an environmental disaster" after acid mine water started flooding the Grootvlei Mine (Groenewald 2010). AMD flowing into the Blesbokspruit was confirmed by DWS water samples (McKay and Milaras 2017, p. 8). The IMC report of 2010 explained the situation in this basin as complicated due to the cessation of pumping at the Grootvlei Mine. The mine had previously maintained the water in the basin over the years of mining at Shaft No. 3 (Coetzee et al. 2010, pp. vi–vii). The Grootvlei Mine was the last operational gold mine on the East Rand at the time until its forced closure in 2011, and prior to its closure was responsible for all pumping (McKay and Milaras 2017, p. 1). Shaft No. 3 was used primarily for pumping water, high-density sludge and settling ponds to treat AMD. Furthermore, ownership of the mine changed several times: in 1998 to Petmin Ltd., in 2000 to Petrex Pty Ltd., then Pamodzi Gold Ltd. in 2006, and to Aurora Empowerment Systems in 2009. The change in ownership was indicative of ongoing financial constraints. The national government had to intervene and subsidised the pumping and treatment of AMD in this basin. Owing to the uncertainty of when pumping would commence, concerns were raised about the water decanting to the surface.

The AMD decanting threat led to the DWS issuing a directive to the mine to pump and treat AMD in 2010. Aurora failed to comply with this duty and criminal charges were laid against it for polluting the Blesbokspruit. In May 2011 the trade union Solidarity launched liquidation actions against Aurora, which resulted in all responsibility falling on the government to deal with AMD problems (McKay and Milaras 2017, p. 10). The DWS applied for authorisation from the Department of Environmental Affairs (DEA) to construct the AMD plant at the Grootvlei Mine's Shaft No. 3 because the situation was granted emergency status by the DWS (Govender, personal interview, 2018). In 2019, directors of Aurora Empowerment Systems made an appearance in the Springs Regional Court, where they faced five counts related to environmental and water transgressions for failing to treat the underground mine water before discharging it into the Blesbokspruit (Mabuza 2019). The trial against the directors was ongoing and was set to take place from 18 to 25 June and 23 to 29 July 2020 (Kubheka 2019; Niselow 2019). On the contrary, the National Prosecution Authority had confirmed that the case was "struck" from the Springs Regional Court, without reasons provided to stop pursuing charges (Sicetsha 2020).

From 2011 until the AMD STT treatment began in 2016 (Tlale, personal interview, 2018a), no pumping took place (Van der Merwe and Lea 2003, p. 25). AMD did not breach and remained below the ECL. "We are dealing with a historical problem and do not have the real culprits and therefore we are dealing with the problem. If we did not make this intervention, our own groundwater would be contaminated, and Gauteng would have been highly affected and going into the future this will be a condition when issuing a water use licence to mining houses" (Then Minister of Water and Sanitation, Nomvula Mokonyane speaking on *SABC News* 2017). According to the Mine Water Management Draft Policy (DWS 2017, p. 2), it was necessary for the DWS to formulate a policy on mine water impacts to "provide the position of the DWS on mine water management, including AMD. To provide measures on protection of water resources from prospective, operational and historical mine activities that have negative quality impacts" (DWS 2017, p. 2) (discussed further in Chap. 7).

According to Govender (personal interview, 2018), the DWS avoided a situation where AMD breached and reached full-scale proportions before the department addressed it and implemented a remedy before it reached crisis stages. "We can actually deal with AMD quite effectively from the start of a mining venture through to its operation and closure, and that's the process. This is obviously supported by various initiatives ranging from how we go out and assess catchments right through to policy imperatives, so it's a broad ranging process," said Govender. Owing to the urgency of the STT, misunderstandings existed between various stakeholders over the legal requirements for the construction of the plant and the STT process (explained in detail below).

4.4 Construction of the Eastern Basin AMD Treatment Plant and STT Process

In 2010 the IMC report presented a STT for AMD in the eastern basin (Coetzee et al. 2010). The DWS consulted with the government departments concerned, namely, the DEA and the Department of Mineral Resources (DMR), and gave the DWS approval and permission to implement the STT. The Trans-Caledon Tunnel Authority (TCTA) is an agency of the DWS that is responsible for implementing the AMD project in the eastern basin (TCTA 2019) (Fig. 4.2).

Tlale (personal interview, 2018a), AMD Project Manager at the TCTA, explained that a limitation in choosing a site to construct the plant was the fact that they had to find a place that was suitable to meet the engineering requirements to put the pumps at the depth required. Grootvlei Mine's Shaft No. 3 proved to be the best option. Once the engineering requirements were approved, the DWS issued a directive to the TCTA for the plant to be constructed (Figs. 4.3 and 4.4) (Govender, personal communication, 2018; Pillay, personal interview, 2018; *Springs Advertiser* 2015). The construction contract was awarded to CMC/PG Mavundla Engineering Eastern Basin Joint Venture at a contract amount of R956 141 123.68, excluding value-added tax (VAT) and escalation. This contract is also for the pump station, neutralisation facility and related infrastructure to be established at Grootvlei Mine's Shaft No. 3 (Digby Wells Environmental 2014). There is a formal project coordination committee for the construction of the plant and the actual STT, within which the DWS is represented, including Bashan Govender (Mine Water Management Unit representative) (Tlale, personal interview, 2018a). The TCTA and the DWS communicate daily via email on the project. In this way, the information is official and decisions made by the DWS can then be implemented by the TCTA (Tlale, personal interview, 2018a). The implementation of the project has three tiers: (1) the DWS responsible for policy development of the AMD treatment process; (2) the TCTA working as the project manager and (3) AECOM South Africa (Pty) Limited appointed by the TCTA, as the principal consultant on the implementation of the proposed STT

Fig. 4.2 Location of eastern basin AMD treatment plant. (Google Maps 2020b)

measures for a sludge disposal facility and the pipeline associated with the treatment of AMD from the eastern basin (Digby Wells Environmental 2015a, p. 7).

While stakeholders waited in anticipation for the AMD treatment plant to be constructed, the concern was that no EIA existed for the construction of the plant at the Grootvlei Mine Shaft No. 3 site (Blesbokspruit Forum 2018a). For the sake of clarity, the construction of the AMD treatment plant and discharge of water into the Blesbokspruit required one approval process, and the disposal of sludge is seperate and required a different approval process. According to Pillay (personal interview, 2018), the DWS was allowed to proceed with the construction of the plant and the pumping of water, which formed part of one process, but were not allowed to dispose of the sludge until they had a required licence for this listed activity according to National Environmental Management Act (NEMA) regulations. However, according to Govender (personal interview, 2018), no EIA was required for the construction of the

Fig. 4.3 Eastern basin AMD treatment plant. (Photographs by Stan Madden)

plant or the pumping of water because the project was declared a government waterworks and constructed under emergency circumstances. This allowed for construction to begin while the required EIA for the sludge disposal was running. Digby Wells Environmental was the chosen environmental consultant (Govender, personal interview, 2018) and conducted a full EIA for the disposal of sludge at Grootvlei Site 6/L/16 TSF, to be used for a period of 5–8 years (discussed in Sect. 4.5). The DEA approved the completed EIA in 2015 (Digby Wells Environmental 2015a).

However, when the STT began, Grootvlei Site 6/L/16 could not be used because Ergo Mining – who own the reclamation rights to this tailings storage facility (TSF) – indicated that they wanted to use this site sooner than anticipated. They initially agreed to allow the DWS to use the site for the disposal of sludge for a

Fig. 4.4 Eastern basin AMD treatment plant entrance. (Photograph by researcher)

period of 5–8 years, which was the period for which the STT was meant to run, but then indicated that the site could only be used for 2 years. This left the DWS in a predicament, with the STT about to begin and no sludge disposal site available for the required period. Using the site for only 2 years would have led to higher cost implications, involving moving the sludge, and rehabilitating the site after the 2 years (Tlale, personal communication and site visit, 2018b). The option of using the Grootvlei Mine void, where the treatment is taking place, surfaced and became the sludge disposal site, and because no EIA exists for the sludge being disposed of in the mine void, this turned into a pilot project.

The trial pumping in the eastern basin started in August 2016, 6 months prior to the launch of the AMD treatment plant (Govender, personal interview, 2018; Tlale, personal interview, 2018a). The AMD treatment plant for the eastern basin was launched on 17 February 2017 (Fig. 4.5) (Digby Wells Environmental 2014; personal attendance of launch; Mokonyane 2017). At the launch, attendees were taken on a tour of the plant and were shown how the plant operated (see Fig. 4.6).

The eastern basin has the biggest AMD treatment plant in the world and is of "top-class standard" (Tlale, personal interview, 2018a). AMD water is abstracted from Grootvlei Mine's Shaft No. 3 and pumped into the high-density sludge plant (see Fig. 4.3), where the water is neutralised or partially treated with lime[3] (Environomics 2014, p. 28; *Springs Advertiser* 2015). The lime neutralises the water, and this partially treated water is then discharged into the Blesbokspruit (Anticevich 2014; Ratau 2016, p. 11), with the sludge being disposed into a disposal site – the mine void.

[3] Part of the process involves adding lime to increase the pH of the water and then "blowing oxygen into the water to oxidise the iron which precipitated, taking with it most of the heavy metals" (Environomics 2014, p. 28). The iron is then separated and disposed of in tailings storage facilities and the water discharged into the nearest watercourse that can accommodate the sum of water.

Fig. 4.5 Minister of Water and Sanitation launching the eastern basin AMD treatment plant. (Photograph by researcher)

Fig. 4.6 Guests taken on a tour of the eastern basin AMD treatment plant at the launch. (Photograph by researcher)

Fig. 4.7 Silos filled with lime. (Photograph by researcher)

During the tour of the plant with Tlale, she explained in detail how the plant operated, illustrated below. AMD water is abstracted from Grootvlei Mine's Shaft No. 3 and then placed into silos, which are filled with lime (in powder form) (Fig. 4.7). The water is then transferred through three train reactors for neutralisation (Tlale, personal communication and site visit, 2018b; Digby Wells Environmental 2015a, p. 30). The proportion of lime required for a specific day is electronically controlled, depending on the pH of the AMD water that comes out of the shaft. The frequency at which the water needs to be treated is then determined, such as how much lime is required and how regularly it needs to be added to that portion of acidic water.

After being treated with lime, the water is exposed to oxygen (Fig. 4.8). As the water is pumped, the iron in the water when in contact with oxygen becomes iron oxide and turns red, which is where the colour of the water starts forming. The sludge is the finished product (Figs. 4.9 and 4.10).

The treated water is then pumped to one of three thickener or settling tanks (Fig. 4.11). From there the partially treated water flows into what is called *a holding tank* before it is discharged into the Blesbokspruit. The sludge from this process is recycled and pumped into sludge pumps (Fig. 4.12); the sludge is then divided, with some disposed into the mine void and some reused for the treatment process. When the STT process started, the sludge used was channelled straight into the mine void, but now there are boreholes which are part of the mine void; three boreholes are used for the sludge (Tlale, personal communication and site visit, 2018b).

Fig. 4.8 Process of water being transferred and exposed to oxygen. (Photograph by researcher)

Fig. 4.9 Sludge. (Photographs by researcher)

Fig. 4.10 Sophia Tlale, AMD Project Manager at the TCTA, explaining the process. (Photographs by researcher)

Water that is acidic has a pH lower than 7, and a normal pH range in surface water systems is 6.5–8.5 (Dirisu and Mafiana 2016; *Daily Star* 2015). Therefore, through this STT process, the pH of the water is increased and would remain much lower if no treatment takes place at all. The AMD treatment plant has the capacity

Fig. 4.11 Settling tanks. (Photographs by researcher)

Fig. 4.12 Sludge pumps. (Photograph by researcher)

to treat up to 110 Mℓ of water per day but has been set to pump at a rate of 80 Mℓ/day (Blesbokspruit Forum 2018a, b, 2019b). The reason for the treatment being 'short term' is that the treatment process is only able to remove the heavy metals, with high sulphate levels remaining. High sulphate levels worsen the quality of the water in the Blesbokspruit receiving the discharged water, which can eventually increase the pollution in the Vaal River since the Blesbokspruit is a tributary of the Vaal (Digby Well Environmental 2015a, p. 7).

4.5 EIA Process for the Selection of a Sludge Disposal Site

As indicated above, in 2015 the DEA approved the final EIA for the disposal of sludge at Grootvlei Site 6/L/16 (Digby Wells Environmental 2015a). However, owing to the predicament with this site, it could not be used and, instead, the mine void at Grootvlei Mine's Shaft No. 3 was chosen as the site to dispose of the sludge. The purpose of the section is to explain why this happened, the process involved in drafting the EIA for the sludge disposal site and the legal requirements of an EIA. This background is important to understand how the various perceptions of the short-term AMD treatment – and water quality – were influenced by different views stemming from the EIA process (discussed in Chap. 5).

Digby Wells Environmental, appointed as the environmental consultant for the Grootvlei Site 6/L/16 sludge disposal site, had to conduct the required regulatory environmental authorisation process. The EIA process conducted was specifically limited to the construction of a sludge disposal facility and associated pipelines for the transfer of sludge and return water between the eastern basin AMD treatment plant (situated at Grootvlei Mine's Shaft No. 3) and the proposed sludge disposal facility at Grootvlei Site 6/L/16 TSF (Digby Wells Environmental 2015a, p. ii). The terms and conditions stipulated in the NEMA and the National Water Act (NWA) were used to conduct the research required for the EIA report and approval of this sludge disposal site. The sludge classification is type three; this means that the nature of the waste is due to elevated concentrations of arsenic, cadmium, manganese, nickel and antimony (Digby Wells Environmental 2015a, p. 4). This type of waste, according to the National Environmental Management Waste Act 59 of 2008, needs to be disposed of within a Class C[4] landfill. Two possible sites were narrowed down and assessed by Digby Wells as possible sludge disposal sites, namely, the Grootvlei Site 6/L/16, situated about 1 km from Grootvlei Mine's Shaft No. 3, and the Largo site, which is located on the eastern banks of the Blesbokspruit and is about 2 km from Grootvlei Mine's Shaft No. 3 (Digby Wells Environmental 2015a, p. 4). Each site had a planned pipeline route from the site to the Blesbokspruit of no more than 300 mm in diameter and with a flow rate of about 21 ℓ/s, implying that both sites were found useful for the disposal of sludge. Each site was assessed for suitability in terms of a range of factors, such as environmental and socio-economic impacts. The outcome of the assessment led to the Grootvlei Site 6/L/16 being selected for the sludge disposal, with a final EIA then conducted for this site.

The main purpose of an EIA is to provide a full description of the planned activity, including a "description of the property on which the activity is to be undertaken and the location of the activity on the property and the activity to be undertaken at the exact coordinates, a description of the environment that may be affected by the activity and the manner in which the physical, biological, social and economic and

[4] A Class C landfill means there are certain requirements in order to dispose of this waste, namely containment barriers of landfills for the disposal of waste must comply with minimum engineering design requirements (RSA 2008, p. 61).

cultural aspects of the environment may be affected by this activity" (Digby Wells Environmental 2015a, p.8).

To obtain authorisation to use a site, three phases are required (Digby Wells Environmental 2015a, p. 3):

- The scoping phase: Interested and affected parties (IAPs) are consulted, information is shared, and environmental concerns are identified and alternative suggestions made. A draft scoping report is then made available for public comment and the final report is then sent to the DEA.
- The EIA phase: Potential impacts and alternatives are discussed with the relevant stakeholders and assessed, compilation of an Environmental Management Programme and a draft EIA report, a public participation process whereby members of the public comment on this draft report, and the final report is submitted to the DEA.
- Environmental authorisation phase: The DEA makes a decision on whether the EIA is approved, the IAPs are informed and they have another opportunity to appeal the decision.

In May 2014 the AMD project was declared a requirement in terms of NEMA to reduce the environmental impact and the EIA process started. The draft scoping report was available for public comment from October 2014 until November 2014 (Digby Wells Environmental 2015b), and the DEA received the final scoping report in late November 2014. The DEA was satisfied with the report as the minimum requirements had been met in terms of the EIA Regulations of 2010. The EIA process was approved with the following requirements to be included in the report: details of future plans for the site and infrastructure after decommissioning in 20–30 years; possibility of upgrading the proposed infrastructure; total footprint on the proposed project; proof of a water use licence if required; possible impacts and effects on vegetation and the industrial area; and how the project benefits the local community (Digby Wells Environmental 2015c). The draft EIA report was available from May 2015 for public (all IAPs) comment for 40 days. The final EIA report was made available for public comments from June to July 2015 (Digby Wells Environmental 2015c). Thereafter, the DEA approved the final EIA. Owing to this approved site for sludge disposal no longer being available once the treatment plant was built, the mine void became the chosen option. Pillay (personal interview, 2018) explained "they are partially treating the water and sending the sludge back down under, which was not in the EIA," and yet the purpose of the EIA was also to assess the impact of the sludge underground.

The DWS Minister had authorised the project to undertake emergency measures to deposit the sludge into an open shaft (mine void), because the plant was ready to start treating the water, but the owner's unexpected need of the approved EIA site would have delayed the treatment and caused AMD to breach (Tlale, personal communication and site visit, 2018b). To avoid AMD breaching, the sludge disposal site project was granted an emergency status and the DWS was exempted from conducting an EIA for the disposal of sludge in the mine void. According to Govender (personal interview, 2018), the use of the mine void was an unforeseen option; the

DWS approached the DEA and the DMR, and both departments conceded that under the circumstances the mine void could be used to deposit the sludge and was considered a pilot project. Tlale (personal communication and site visit, 2018b) indicated that when a project is given emergency status, no EIA is required as stipulated in the NWA (RSA 1998, p. 52). Chapter 11 Section 110 of the NWA explains that consultation and an EIA are required before constructing a waterworks project. However, Subsection 2 states that this does not apply to waterworks that are constructed under emergency circumstances (discussed in more detail in Sect. 7.3.4).

4.6 Conclusion

The aim of this chapter was to provide an overview of the nature of AMD in the eastern basin, as a result of which an urgent need for water treatment was created. It was important to illustrate the STT process, which included the extensive EIA process for the disposal of sludge in the Grootvlei Site 6/L/16. The fact that this site was not used when the treatment began due to a last-minute predicament with the site forms a significant part of this research as to how water quality is socially constructed. These social constructions of water quality can stem from uncertainty about the STT of AMD. This uncertainty arises from the fact that the mine void was used as a last option to dispose of the sludge and is considered a pilot project. This means that the DWS was exempted from conducting an EIA, implying that the risks of using the mine void are unknown, leading to ongoing and increasing concerns and growing negativity surrounding the treatment. The government had many years in which to prepare and address the issue. The IMC report was available in 2010, which already indicated the severity of AMD if not addressed. The ongoing court cases with the directors of Aurora Empowerment Systems showed the impact of non-compliance of a water use licence and the effect it had on the Blesbokspruit. All these factors have added to the way the public feel about the STT of AMD.

The findings of the research are presented in the next three chapters. Chapter 5 presents the stakeholders' comments extracted from the EIA documents. These comments are referred to in combination with in-depth key individual interviews that were conducted for this research to present the perceptions of the AMD STT, which is the main purpose of this chapter; in other words, Chap. 5 indicates *what* the perceptions of water quality in the Blesbokspruit are in the context of AMD and its STT. These perceptions are then used to explain the factors (i.e. the *why*) that influence social constructions of water quality of the Blesbokspruit (discussed in Chap. 6). Chapter 7 explains *how* these social constructions of the water quality stem from power relations; showing how water is intrinsically social and because of its social nature, leads to conflict over how water is used, governed and managed by the various stakeholders.

References

Adler, R., Claasen, M., Godfrey, L. & Turton, A. 2007. Water, mining and waste: An historical and economic perspective on conflict management in South Africa. *The Economics of Peace and Security Journal*, 2(2):33–41.

Ambani, AE. 2013. *Long-term assessment of the surface water quality in the Blesbokspruit Ramsar wetland*. Unpublished MSc dissertation. Johannesburg: University of Johannesburg. Available at: https://ujcontent.uj.ac.za/vital/access/manager/Repository/uj:11619. Accessed on: 21 June 2019.

Anticevich, A. 2014. No, no to this sludge dam: Residents are fighting for the preservation of Blesbokspruit, their properties. *Springs Advertiser,* 28 July. Available at: https://springsadvertiser.co.za/96610/sludge-dam/. Accessed on: 10 January 2018.

Bega, S. 2021. State halts in R10bn long-term plan to fully treat acid mine water. *Mail & Guardian*, 12 June 2021. Available at: https://mg.co.za/environment/2021-06-12-state-halts-its-r10bn-long-term-plan-to-fully-treat-acid-mine-water/. Accessed on 5 October 2021.

Blesbokspruit Forum. 2018a. Minutes of Blesbokspruit Forum meeting and personal attendance, 8 February 2018.

Blesbokspruit Forum. 2018b. Minutes of Blesbokspruit Forum meeting and personal attendance, 11 May 2018.

Blesbokspruit Forum. 2019b. Minutes of Blesbokspruit Forum meeting, 9 May 2019.

Blesbokspruit Forum. 2020. Minutes of Blesbokspruit Forum meeting, 6 February 2020.

Bosman, C. 2018. Consultant for Environmental health services. Personal interview, 12 April. Pretoria, South Africa.

Business Report. 2011. Acid mine water threatens Gold Reef City. *Business Report,* 24 February. Available at: http://www.iol.co.za/business-report/economy/acid-mine-water-threatens-gold-reef-city-1031866. Accessed on: 4 April 2017.

Coetzee, H., Hobbs, PJ., Burgess, JE., Thomas, A. & Keet, M. (eds.) 2010. *Mine water management in the Witwatersrand Gold Fields with special emphasis on acid mine drainage* (Report to the Inter-Ministerial Committee on Acid Mine Drainage). Pretoria: Department of Water Affairs and Forestry. Available at: http://www.dwaf.gov.za/Documents/ACIDReport.pdf. Accessed on: 24 February 2012.

Crowley, K. 2015. Court finds Mandela, Zuma heirs liable for Aurora decay. *Mail and Guardian*, 25 June. Available at: https://mg.co.za/article/2015-06-25-mandela-zuma-heirs-liable-for-sa-mine-assets-decay. Accessed on: 7 October 2019.

Daily Star. 2015. What should be the pH value of drinking water? *Daily Star*, 6 September. Available at: https://www.thedailystar.net/health/what-should-be-the-ph-value-drinking-water-138382. Accessed on: 17 March 2020.

Department of Water and Sanitation (DWS). 2017. *Mine Water Management: Policy Position Draft for External Consultation and Discussion*. Gazette No. 658. Available at: https://www.greengazette.co.za/notices/national-water-act-36-1998-mine-water-management-policy-position-draft-for-external-consultation-and-discussion_20170707-GGN-40966-00658.pdf. Accessed on: 18 October 2018.

Digby Wells Environmental. 2014. *Environmental impact assessment for the construction of the proposed sludge disposal facility and pipeline associated with the treatment of acid mine drainage from the eastern basin of the Witwatersrand, Gauteng: Full comment and response report*. Johannesburg: Digby Wells and Associates.

Digby Wells Environmental. 2015a. *Construction and operation of the proposed sludge disposal facility and pipelines associated with the treatment of acid mine drainage in the eastern basin of the Witwatersrand, Gauteng: Final environmental impact assessment report*. Johannesburg: Digby Wells and Associates.

Digby Wells Environmental. 2015b. *Notification of application for environmental authorisation and waste management licence for the proposed sludge disposal facility and pipeline associated with the treatment of acid mine drainage from the eastern basin of Witwatersrand gold fields, Gauteng Province: Availability of draft environmental impact assessment report for public comment.* Public Letter April 2015.

Digby Wells Environmental. 2015c. *Notification of application for environmental authorisation and waste management licence for the proposed sludge disposal facility and pipeline associated with the treatment of acid mine drainage from the eastern basin of Witwatersrand gold fields, Gauteng Province: Availability of final environmental impact assessment report for public comment.* Public Letter June 2015.

Dirisu, C. & Mafiana, M. 2016. Level of pH in drinking water of an oil and gas producing community and perceived biological and health implications. *European Journal of Basic and Applied Sciences*, (3):3. Available at: https://www.researchgate.net/publication/332012834_level_of_ph_in_drinking_water_of_an_oil_and_gas_producing_community_and_perceived_biological_and_health_implications. Accessed on: 14 august 2020.

Environment. 2016. Acid mine drainage in South Africa. Available at: https://www.environment.co.za/poisoning-carcinogens-heavy-metals-mining/acid-minedrainage- amd-south-africa.html. Accessed on: 10 February 2020.

Environomics. 2014. *Gauteng environmental management framework report.* Johannesburg: Gauteng Province, Agriculture and Rural Development. Available at: http://www.gauteng.gov.za/government/departments/agriculture-and-rural-development/Documents/Gauteng%20Provincial%20Environmental%20Management%20Framework/2014%20GPEMF%20Section%20A%20Introduction%20and%20Status%20Quo.pdf. Accessed on: 20 June 2019.

Evans, S. 2014. Jacob Zuma, Khulubuse and the missing millions. *Mail and Guardian,* 16 May. Available at: https://mg.co.za/article/2014-05-15-jacob-zuma-khulubuse-and-the-missing-millions. Accessed on: 7 October 2019.

Fin24. 2019. Khulubuse Zuma, Zondwa Mandela in court on Grootvlei Mine charges. *Fin24,* 28 May. Available at: https://www.fin24.com/Companies/Mining/khulubuse-zuma-zondwa-mandela-in-court-on-grootvlei-mine-charges-20190528. Accessed on: 10 June 2019.

Google Maps. 2020a. Location of AMD treatment in the Witwatersrand. Available at: https://www.google.com/maps/search/amd+treatment+plant/@-26.2403807,27.9721857,80262m/data=!3m1!1e3. Accessed on: 9 February 2020.

Google Maps. 2020b. Location of eastern basin AMD treatment. Available at: https://www.google.com/maps/search/eastern+basin+amd+treatment+plant/@-26.2913028,28.4492051,20057m/data=!3m1!1e3. Accessed on: 9 February 2020.

Govender, B. 2018. Chief directorate Mine water management unit: Department of Water and Sanitation. Personal interview, 1 March. Pretoria, South Africa.

Groenewald, Y. 2010. Aurora mine's toxic water crisis. *Mail & Guardian,* 11 June. Available at: https://mg.co.za/article/2010-06-11-aurora-mines-toxic-water-crisis. Accessed on: 7 October 2019.

Guedes, G. 2010. Acid mine drainage still in focus. In *Mining: An indepth discussion of mining issues in S.A* (2010 edition). Edited by Chamber of Mines of South Africa. Cape Town: Nelida Publishing: 60–69. Available at: http://www.bullion.org.za/documents/mining-november-2010.pdF. Accessed on: 14 April 2013.

Hobbs, P., Oelofse, SHH. & Rascher, J. 2008. Management of environmental impacts from coal mining in the upper Olifants River catchment as a function of age and scale. *Water Resources Development*, 24(3):417–431. Available at: http://www.orangesenqurak.com/UserFiles/File/OtherV2/Management%20of%20Environmental%20Impacts%20from%20Coal%20Mining%20Hobbs%20et%20al.%202010.pdf. Accessed on: 14 April 2012.

Holden, R. 2019. Business Analyst: TCTA. Personal Communication, 12 March.

Kings, S. 2016. The victims of the Aurora debacle continue to do what they know best: Mining. *Mail & Guardian,* 9 September. Available at: https://mg.co.za/article/2016-09-09-00-aurora-ZamaZamas-out-of-options-to-find-other-ways-to-make-a-living#comment_thread. Accessed on: 7 October 2018.

Kubheka, T. 2019. Former Aurora Mine directors expected back in court. *Eyewitness News*, 7 August. Available at: https://ewn.co.za/2019/08/07/aurora-mine-former-directors-due-in-springs-court. Accessed on: 17 March 2020.

Labuschagne, DC. 2015. A 3D geological model for the East Rand Basin, South Africa. MSc dissertation. Potchefstroom: North West University. Available at: https://dspace.nwu.ac.za/bitstream/ handle/10394/15351/Labuschagne_DC.pdf?sequence=1&isAllowed=y. Accessed on: 4 October 2018.

Liefferink, M. 2012. Environmental risks and hazards pertaining to AMD and radioactivity within the Witwatersrand gold fields. PowerPoint Presentation. Sandton: Federation for a Sustainable Environment.

Liefferink, M. 2016. CEO: Federation for a Sustainable Environment. Personal interview, 21 November. Bryanston, South Africa.

Mabuza, E. 2019. Khulubuse Zuma, Zondwa Mandela in court over Aurora acid mine drainage. *Timeslive*, 28 May. Available at: https://www.timeslive.co.za/news/south-africa/2019-05-28-khulubuse-zuma-zondwa-mandela-in-court-over-aurora-acid-mine-drainage/. Accessed on: 7 October 2019.

Mail & Guardian. 2010. Aurora's East Rand: The next target? *Mail & Guardian*, 26 November. Available at: https://mg.co.za/article/2010-11-26-auroras-east-rand-the-next-target/. Accessed on: 14 August 2020.

Mail & Guardian. 2011. Aurora told to pay up or close down. *Mail &Guardian* Online, 20 May. Available at: https://mg.co.za/article/2011-05-20-aurora-told-to-pay-up-or-close-down. Accessed on: 15 August 2015.

McKay, TJM. & Milaras, M. 2017. Public lies, private and the forced closure of Grootvlei Mine, South Africa. *The Journal for Transdisciplinary Research in Southern Africa*, 12(1). Available at: https://td-sa.net/index.php/td/article/view/347/399. Accessed on: 12 June 2019.

Mokonyane, N. 2017. Statement by the Minister of Water and Sanitation at the launch of the Eastern Basin long-term treatment plant [Personal attendance], 17 February.

Mouton, S. 2018. From crisis to opportunity: Lessons from Cape Town water shortage. Available at: https://www.wits.ac.za/news/latest-news/research-news/2018/2018-05/from-crisis-to-opportunity.html. Accessed on: 28 May 2018.

Mpofu, C., Morodi, TJ. & Hattingh, JP. 2018. Governance and socio-political issues in management of acid mine drainage in South Africa. *Water Policy*, 20:77–89. Available at: https://iwaponline.com/wp/article/20/1/77/38136/Governance-and-socio-political-issues-in. Accessed on: 1 June 2020.

Naidoo, S. 2017. *Acid mine drainage in South Africa: Development actors, policy impacts and broader implications*. Switzerland: Springer Nature.

Niselow, T. 2019. Trial date finally set for Khulubuse Zuma, Zondwa Mandela in water pollution case. *Fin24*, 22 November. Available at: https://www.fin24.com/Companies/trial-date-finally-set-for-khulubuse-zuma-zondwa-mandela-in-water-pollution-case-20191122. Accessed on: 21 May 2020.

Parliamentary Monitoring Group (PMG). 2010. *Aurora Grootvlei Mine issues: Department of Labour report*. Cape Town: PMG. 24 August. Available at: https://pmg.org.za/committee-meeting/11913/. Accessed on: 7 October 2019.

Parris, K. 2010. *Water in agriculture: Improving resource management*. Paris: Trade and Agriculture Directorate, OECD. Available at: http://www.oecdobserver.org/news/archivestory.php/aid/3217/Water_in_agriculture:_Improving_resource_management.html. Accessed on: 11 February 2014.

Pillay, M. 2018. Business Development Executive: Digby Wells Environmental. Personal interview, 16 January. Bryanston, South Africa.

Ratau, S. 2016. Opposing legacy of acid mine drainage. *The Star*, 22 August.

Republic of South Africa (RSA). 1998. *National Water Act, no 36 of 1998*. Pretoria: Government Printer. Available at: http://www.dwaf.gov.za/Documents/Legislature/nw_act/NWA.pdf. Accessed on: 14 April 2013.

Republic of South Africa (RSA). 2008. *National Environmental Management: Waste Act no 59 of 2008*. Pretoria: Government Printer. Available at: https://www.gov.za/sites/default/files/gcis_document/201409/35572gen615.pdf. Accessed on: 10 March 2020.

SAnews. 2017. Draft Mine Water Policy out for public comment. Available at: https://www.sanews.gov.za/south-africa/draft-mine-water-policy-out-public-comment. Accessed on: 24 January 2020.

SABC News. 2017. Acid mine drainage treatment plant launched in the East Rand. *SABC News,* 17 February. YouTube online. Available at: https://www.youtube.com/watch?v=hGF4PHUu4HU. Accessed on: 2 September 2018.

Scott, R. 1995. *Flooding of the Central and East Rand gold mines: An investigation into controls over the inflow rate, water quality and predicted impacts of flooded mines* (Water Research Report No. 486/1/95). Pretoria: Water Research Commission. Available at: http://www.wrc.org.za/wp-content/uploads/mdocs/486-1-95.pdf. Accessed on: 11 June 2019.

Sicetsha, A. 2020. Khulubuse Zuma, Zondwa Mandela scot-free after NPA drops charges. *The South African,* 26 February. Available at: https://www.thesouthafrican.com/news/khulubuse-zuma-zondwa-mandela-scot-free-after-npa-drops-charges/. Accessed on: 10 October 2020.

Springs Advertiser. 2015. Plant will be operational in a year. *Springs Advertiser*, 18 February. Available at: https://springsadvertiser.co.za/113292/plant-will-be-operational-in-a-year/. Accessed on: 10 January 2018d

Trans-Caledon Tunnel Authority (TCTA). 2019. TCTA website. Available at: https://www.tcta.co.za/. Accessed on: 14 August 2020.

Thorius, T. 2004. *The effect of Grootvlei mine water on the Blesbokspruit*. MSc dissertation on Environmental Management. Johannesburg: University of Johannesburg. Available at: https://ujcontent.uj.ac.za/vital/access/manager/Repository/uj:14868. Accessed on: 8 July 2019.

Tlale, S. 2018a. AMD project manager: TCTA. Personal interview, 9 March. Centurion, South Africa.

Tlale, S. 2018b. AMD project manager: TCTA. Personal communication and AMD Treatment Plant Site Visit, 09 April 2018. Springs, South Africa.

Turton, A., Schultz, C., Buckle, H., Kgomongoe, M., Malungani, T. & Drackner, M. 2006. Gold scorched earth and water: The hydropolitics of Johannesburg. *Water Resources Development*, 22(2):313–335.

Turton, A. 2015. Sitting on the horns of a dilemma: Water as a strategic resource in South Africa. *@Liberty*, 6(22). November 2015. Available at: http://irr.org.za/reports-and-publications/atLiberty/files/liberty-2013-sitting-on-the-horns-of-a-dilemma-2013-water-as-a-strategic-resource-in-south-africa. Accessed on: 18 March 2016.

Turton, A. 2018. Mine voids. Personal communication, 18 October 2018.

Van der Merwe, W. & Lea, I. 2003. Towards sustainable mine water treatment at Grootvlei Mine. In *Mine Water and the Environment. Proceedings of the 8th International Mine Water Association Congress*. Johannesburg, South Africa.

Whittles, G. 2016. Aurora mine liquidators on the hunt for Khulubuse Zumas hidden riches. *Mail & Guardian* 9 September. Available at: https://mg.co.za/article/2016-09-09-00-aurora-mine-liquidators-on-the-hunt-for-khulubuse-zumas-hidden-riches. Accessed on: 7 October 2019.

Chapter 5
Perceptions of Water Quality in the Context of the Short-Term Treatment of Acid Mine Drainage in the Eastern Basin

Abstract This chapter specifically explores the perceptions of stakeholders about the water quality of the Blesbokspruit, in the context of acid mine drainage (AMD) and its short-term treatment (STT). The purpose of the STT is to reduce the negative impact of AMD on the water quality. However, because the implementation was delayed, and changes were made over how it was implemented, it created various perceptions of the STT regarding what these changes meant for the water quality. These perceptions are further attributed to how people viewed the management of the Blesbokspruit. Perceptions about the environmental impact assessment (EIA) for the sludge disposal site constantly came up in conversations about the water quality of the Blesbokspruit. Therefore, an in-depth review of the documents that formed part of the EIA process for Grootvlei Site 6/L/16 was conducted, given that it was a vital part in how perceptions of the STT and water quality were formed. This chapter draws on various comments raised by the interested and affected parties (IAPs) during the EIA process. This information is used in combination with in-depth interviews that were conducted for this research to present the perceptions of AMD STT and of water quality. This includes the different perceptions and views that stakeholders and key individuals hold who are directly linked to the AMD treatment process and management of the Blesbokspruit catchment. The key individuals' views of the water quality of the Blesbokspruit are provided. Then various stakeholders' and key individuals' perceptions of the extent of AMD in the eastern basin, perceptions before the STT began, as well as their views on the actual STT process, including the discharge of water (volumes and sulphate levels) and the sludge disposal, are presented. The relevance of noting these views and perceptions is to provide a background to how the social constructions of the water quality in the Blesbokspruit formed.

Keywords Perceptions · Water quality · Short-term treatment · Acid mine drainage

S. Naidoo, *Social Constructions of Water Quality in South Africa*,
https://doi.org/10.1007/978-3-030-98237-9_5

5.1 Introduction

This chapter specifically explores the perceptions of stakeholders about the water quality of the Blesbokspruit, in the context of acid mine drainage (AMD) and its short-term treatment (STT). The purpose of the STT is to reduce the negative impact of AMD on the water quality. However, because the implementation was delayed, and changes were made over how it was implemented, it created various perceptions of the STT regarding what these changes meant for the water quality. These perceptions are further attributed to how people viewed the management of the Blesbokspruit (discussed in Chap. 6).

It is important to note that reference to the environmental impact assessment (EIA) process refers to the approved EIA that was conducted for the sludge to be disposed at Grootvlei Site 6/L/16 tailings storage facility (TSF), and not the mine void that eventually became part of the STT. Further, the AMD treatment plant at Grootvlei Mine's Shaft No. 3 and the discharge of neutralised water into the Blesbokspruit did not require an EIA, and the use of the mine void was exempted from an EIA, as indicated in the previous chapter.

Perceptions about the EIA for the sludge disposal site constantly came up in conversations about the water quality of the Blesbokspruit. It was therefore crucial for the research to conduct an in-depth review of the documents that formed part of the EIA process for Grootvlei Site 6/L/16, given that it was a vital part in how perceptions of the STT and water quality were formed. This chapter draws on comments in the EIA documents from the interested and affected parties (IAPs) and is used in combination with in-depth interviews that were conducted for this research to present the perceptions of AMD STT and water quality. This includes the different perceptions and views that stakeholders and key individuals hold who are directly linked to the AMD process and management of the Blesbokspruit catchment. The key individuals' views of the water quality of the Blesbokspruit are provided. Then various stakeholders' and key individuals' perceptions of the extent of AMD in the eastern basin, perceptions before the STT began, as well as their views on the actual STT process, including the discharge of water (volumes and sulphate levels) and the sludge disposal, are presented. The relevance of noting these views and perceptions is to provide a background to how the social constructions of the water quality in the Blesbokspruit formed.

5.2 How Key Individuals Understand the Water Quality of the Blesbokspruit

> *What do you need the water for in order to say the quality is good.* (Maurizi, personal interview, 2018)

This section sets out how the key individuals understand the water quality of the Blesbokspruit in terms of 'fitness for use', based on how they view the quality, how the quality is measured and why it is done this way. The concept 'fitness for use' is central to water quality management in South Africa. According to DWAF (1996, p. 4), "The fitness for use of water is a judgement of how suitable the quality of water is for its intended use or for protecting the health of aquatic ecosystems." The discussions in this section present information that amounts to an understanding that the influences and impacts on the value and quality of water on perceptions, depending on the use for water, and ultimately on social constructions, do not follow a linear causal route, but work in multi-linear directions.

People have various interests in the water of the Blesbokspruit and its quality, which can be understood by how it is managed and by how they measure water quality to indicate their institutional interest in water. Constructs come from how one views or understands reality, and therefore, how one perceives water quality is based on one's understanding of how people attribute meaning to water and its quality. The way in which individuals view water begins to indicate their social relationship with water. Individual knowledge, views and experience of poor-quality water stem from the known factors already impacting on the Blesbokspruit, such as the history of mining, coal mining, flows and volumes of water, technology and processes used, and access to information and communication and vested interests (discussed in Chap. 6). Such factors could have influenced the perceptions of AMD STT based on the unknown impacts or that it could further exacerbate the known pollution problems. This was due to the fact that no EIA was conducted to identify what possible impacts could occur. Such factors suggest how social constructions of the water quality are formed, through peoples' interactions with water. This section is placed at the start of the chapter to show that how these individuals understand water quality is based on how they perceive the STT of AMD. It is important to note that in this catchment water quality is managed according to a set of In-stream Water Quality Guidelines (Reservoir website 2021). The way some stakeholders and key individuals understand water quality is based on these water quality guidelines.[1]

From a social perspective, if one is talking about water quality one would not refer primarily to scientific measurements, but it would be purely from that individual's point of view or understanding of information. Such viewpoints depend on how one defines water quality and is understood as such by what is present in the water (Joshua, GDARD representative, personal interview, 2019) and, further, what the use of the water is. This view of water quality starts with the visual appeal, and Jonhasi, who is an aquatic ecologist at the GDARD, argued that this determines whether (s)he can/will drink it. For a layperson, good water quality means it can be

[1] A water quality guideline is a set of information provided for a specific water quality constituent. It consists of the water quality criteria, including the Target Water Quality Range, for that constituent together with other support information such as the occurrence of the constituent in the aquatic environment, the norms used to assess its effects on water uses, how these effects may be mitigated, such as identifying possible treatment options. The South African Water Quality Guidelines include guidelines for domestic, recreational, industrial and agricultural water uses, the protection of aquatic ecosystems and the protection of the health and integrity of aquatic ecosystems and the protection of the marine environment (DWA 1996, pp. 2–3).

used domestically or for consumption. If it is unsuitable for human consumption, then the quality is regarded as poor (Mashau and Jonhasi, GDARD representatives; personal interview, 2019). Maluleke (GDARD representative, personal interview, 2019), an employee of the GDARD who is based at the Marievale Bird Sanctuary, proposed that if the water looked safe to consume, a person on the street would drink it, meaning the visual appeal influenced the perception of quality.

Similarly, Fourie, an aquatic ecologist who consults for the City of Ekurhuleni (CoE) and who is responsible for conducting their water quality assessments for the Blesbokspruit catchment, stated that if the water looked poor in quality, he would not drink it and he would not come into contact with it. According to Maluleke, if one grows up in a village, has cattle and has not drunk water in hours, whatever water is available, even if it is from a little stream with minimal movement (as long as the colour is clear and the ground is visible), one would drink the water and so will the cattle. Mashau, the Project Administrator for the Blesbokspruit Ramsar site at the GDARD,[2] added that in such cases people with local knowledge will know that others (village people) were also drinking this water. However, if it is unsafe for consumption, they will already be aware that it cannot be consumed and they would not allow their cattle to drink it either. A further description by individuals with local knowledge of the water quality of the Blesbokspruit is illustrated by van der Merwe (personal interview, 2018) and Madden (personal interview, 2018a). Van der Merwe is a representative on the Blesbokspruit Forum and the Chairperson of the Blesbokspruit Trust, and Madden[3] is a well-known environmental activist in Springs; he placed Marievale Bird Sanctuary on the map by helping to ensure that larger portions of the land remain protected for the bird sanctuary, therefore he is also linked to the tourist sector. Both are also community members and stated that even though there are no apparent health impacts from consuming the water of the Blesbokspruit, they would not drink it. There are many people who are forced to drink it (e.g. people living in the informal settlements), and "for us as custodians, it's not right to accept this" (Madden, personal interview, 2018a).

While other stakeholders also refer to the visual appeal, they did so for different reasons. If water is used for drinking purposes or domestic purposes, then the visual appeal must be convincing to enable them to consume it without having to measure it (through sampling), explained Jonhasi. However, the quality of the water is sometimes too poor even for domestic use or drinking, but it can be used for building. Those who consider water quality from a scientific perspective will include elements (i.e. chemicals, pollution) that exist in the water. Therefore, it is difficult to obtain a common perspective on what water quality is because it depends on what one needs it for.

Maurizi is a Springs resident and Water Quality Manager at the CoE. To her the word 'quality' is more inclined to be viewed from a chemistry perspective, given

---

[2] B. Mashau chairs the Gauteng Wetland Forum and is a GDARD representative in the Blesbokspruit Forum.

[3] S. Madden also worked in the mining industry in the East Rand for over 40 years.

her chemistry background and her role at the CoE. She views water quality in both its domestic use and in terms of the set guidelines required by the CoE who are guided by the Blesbokspruit Water Quality Guidelines (which are not the same for all rivers). When monitoring the water, these guidelines are used to determine whether the measured water quality is good or bad (Maurizi, personal interview, 2018). Maurizi and Mashau (GDARD representative, personal interview, 2019) suggested that professionals would look at water quality in terms of the chemical components. Fourie (personal interview, 2019) terms it the "physiochemical habitat," because the quality is determined once it is scientifically measured, rather than from a visual appeal. Both Fourie and Maurizi suggest that measuring water quality from a scientific perspective is not the only way, but it is based on their professional backgrounds. Further, it is important to ask what should be measured, and therefore it is difficult to measure based on each user's need (Maurizi, personal interview, 2018). This makes it difficult for individuals to have similar views on what the quality of the water is because of the different indicators they look for to determine if the habitat is suitable for their needs. The need and use for water are a decisive factor in how the water quality is viewed and what indicators that person uses to scientifically measure and view the water quality.

In addition to the indicators used to measure water quality scientifically, when one talks about absolute quality, there are specifications that stipulate certain chemicals not exceeding specific levels (Jonhasi, GDARD Representative, personal interview, 2019). These specifications suggest that the pH and acidity must be at specific levels. If one is looking at the absolute water quality, then it has to meet the prescribed quality requirements for the river as per CoE water quality guidelines. A purpose of the Blesbokspruit Forum is to report whether the river water quality is meeting the In-stream Water Quality Guidelines, such as by checking the amount of salts in the water (Joshua, GDARD representative, personal interview, 2019). Such criteria are used as variables to determine water quality. This is purely from a scientific perspective, and a professional would refer to this description of water quality, but a layperson would not.

These scientific indicators can further determine whether the water is good enough for human consumption and/or the environment. Some look at the quality of water by accounting for the aquatic life. Fourie views the water quality based on his water quality assessments and results and applies this to how he views the water, such as the fish response assessment index. From his assessments, the quality of water is not that bad, considering the history of AMD issues in the Blesbokspruit, where the pH was extremely low at one stage. Furthermore, the nutrient value in the water depends on how urbanised the area is and looking at the discharges into the water from the various users. The river is still providing a habitat for fish species. However, high levels of the *Escherichia coli* (*E. coli*) bacterium and tailings seepage contribute to the varied water quality, but the system tends to balance itself out (Fourie, personal interview, 2019). If the people were taken out of the area, then the water quality would be good; but while there are people living there and using the water, the way the water is managed has to account for all users (Maurizi, personal interview, 2018). It is therefore about maintaining the socio-economic balance;

managing the quality of water while being aware that people are using the water, which means there are going to be negative impacts on the quality. The aforementioned requires maintenance and proper management to minimise the impacts so that the quality is good enough to sustain the river system (Maurizi, personal interview, 2018).

These are the reasons why Maurizi suggests that it is difficult to view water quality from one perspective, because everyone wants something different and is using the water for different reasons. One of the most challenging aspects is taking into account the quality from a microbiological and chemical perspective. When the water is sampled from a microbiological perceptive, the elevation of pollution can be identified. This could be due to wastewater treatment works, for instance, where *E. coli* is evident (through sampling) in the water. According to Fourie (personal interview, 2019), the CoE can then immediately identify whether the poor water quality is due to *E. coli*. This means that the CoE knows who has a licence to discharge water and can either attribute the pollution to those licence holders or investigate whether there is an illegal discharge of water by a user without a licence. When the water is sampled from a chemical perspective, it is more difficult to pinpoint the source of pollution due to the various users of the water who could be responsible, such as the various mines who have water use licences to discharge into the river.

Due to the various users of the Blesbokspruit, the water quality is hard to quantify, given the fact that there are so many unknown impacts on the water (Maurizi, personal interview, 2018). For instance, there are sources of pollution that cannot be accounted for, such as the mine dumps; this source of pollution can contribute to the decline in water quality (Tlale, personal interview, 2018a). The causes that contribute to poor water quality cannot be pinpointed because they are coming from possible illegal uses of the water. Tlale, who is the Environmental Manager at the TCTA and specifically dedicated to the AMD project, is interested in these sources because it can negatively impact on the way people view the discharged neutralised water from the AMD treatment plant and can attribute the discharge to the poor water quality. These perceived negative impacts on the water due to pollution tend to serve as a departure point for varied views of the water quality discussed in the sections below.

The CoE tries to identify the various sources of pollution. This is done through a monitoring programme, whereby minimum standards determined by the CoE's water quality guidelines must be met, and the variables measured are determined and amended based on the needs of the area. The water quality guidelines for the Blesbokspruit are based on a range of factors, indicated in the South African Water Quality Guidelines (Mashau, GDARD representative, personal interview, 2019). The area is highly industrialised, and the CoE will include variables from this perspective. For instance, owing to mining and the AMD treatment, sulphates would be an important variable. Therefore, the CoE assesses the health of the water based on its intended use; based on who interacts with the Blesbokspruit to determine the variables used to measure the water quality. This includes all the known users of the water, and those that hold water use licences to discharge into the Blesbokspruit. Mashau explained that the expectations of water quality in the river differ. If someone wanted to conduct a mining operation, the licence conditions would not be the same as other river systems or the same for other users, such as for wastewater treatment works.

Due to the diverse users, many factors play a role that impact on the water quality (Fourie, personal interview, 2019), and it is not only associated with AMD treatment discharge releases. There are so many other issues, such as seeping from other tailings, also affecting the system. The discharge from the AMD plant is known, and therefore, negative water quality of the river is attributed to this known source. However, industrial discharge, effluent discharge, farming activities and mining activities also need to be considered (Digby Wells Environmental 2015). An analysis of these contributing sources of pollution was considered in the EIA process for the selection of the Grootvlei Site 6/L/16. However, these factors were not assessed for the mine void. This justifies Tlale's concern about the sources of pollution that cannot be identified. Thus, the CoE cannot account for the pollution caused by other sources, other than conducting investigations. More so, conducting specialist studies, which is part of an EIA process, could have benefited the AMD STT project, even if it meant being conducted after the treatment had commenced.

Even though van Niekerk (personal interview, 2018) proposed that the water quality of the Blesbokspruit was better after the launch of the treatment plant, he suggested that one must understand the historical context. Van Niekerk is a consultant at Karan Beef Feedlot whose main task is to analyse the water quality of the Blesbokspruit. He explained that in 1998/9 water quality monitoring by various departments, such as the CoE and Rand Water, began in the Blesbokspruit at various sample points. At this time there was a vast amount of chemical pollution in the Blesbokspruit from the mining industry. The Catchment Forum Management at the time set ideal conditions (in other words, water quality guidelines) for the Blesbokspruit. At that time everything was just about red[4] in the water, according to van Niekerk, indicating the poor water quality (De Jager, personal communication, 2018a). Many industries were involved in the Blesbokspruit Forum at the time; everyone who impacted or could impact on the Blesbokspruit was involved, including the CoE, which created awareness of the water quality for all users of the water (Van der Merwe, personal interview, 2018). Van der Merwe explained that over time the water quality improved significantly, so much so that today – in terms of chemical pollution – the colour is about green.[5] This indicated to him that the water was of better quality. Sulphates are a bit high, and now *E. coli* levels are constantly reported as high (Blesbokspruit Forum 2018a. b. c. d, 2019a, b, c, d). The water quality in the catchment is discussed at the quarterly meetings of the Blesbokspruit Forum. If there are any irregularities present, based on what the acceptable levels are, the sources of pollution are meant to be discussed and investigated by the department concerned.

[4] This means that untreated water from mining was discharged into the river.

[5] The Water Quality Guidelines for the Blesbokspruit catchment measure different variables according to four categories: blue indicates an ideal catchment background; green indicates an acceptable management target; yellow indicates a tolerable interim target and red indicates unacceptable water quality (Reservoir website 2021).

In the latter part of 2018 and early 2019, overall the water quality status of the Blesbokspruit catchment had improved and was reported by representatives of various government departments to be in a better condition (Blesbokspruit Forum 2018c, 2019b). The reason for the improvement was attributed to the high rainfall in the catchment during the summer period of January to March 2019, which had a diluting effect on the system. However, this could also be due to the fact that the sulphate levels were low, because the AMD treatment plant had stopped pumping and discharging water during this time due to a fault in the system (Madden, personal communication, 2018c; Tlale, personal communication, 2019).

Since the STT of AMD began, *E. coli* was reported at unacceptable levels (Blesbokspruit Forum 2017, 2018a, b, c, d, 2019a, b, c, d). This contributed to the poor water quality of the Blesbokspruit. There is the problem of biological pollution in the Blesbokspruit (mainly *E. coli*), and in terms of water quality testing this is probably the only indicator that continuously shows as red (unacceptable levels) on the sample sheets. People have changed the water system so much due to their particular use of the water that they need to ask "what do we want to survive in the river because there are changes in the whole system?" (Maurizi, personal interview, 2018). This means that human-created events that occur in the environment lead to socially produced ways of experiencing and understanding the environment. People have contributed to how the system has changed because of how water is social. The Blesbokspruit used to be a wet and dry system – in winter dry and summer it would be wet – hence, the name *spruit*. However, with urbanisation there is now a constant flow, with wastewater treatment plants and the AMD treatment plant pumping water into the Blesbokspruit all year long. These activities not only play a role in the amount of water but also the water quality. These various changes to the system would bring different views (on the water quality) as some people do not want treated wastewater discharged into the river, "but where else must it go" (Maurizi, personal interview, 2018).

This discussion demonstrated how individuals who are linked to the Blesbokspruit understand water quality. How individuals define and understand water quality depends on their interest in, or interaction with, the water. These views developed from the individual's interest in water, based on their professional background, such as how water quality is measured through the In-Stream Water Quality Guidelines or whether the individual lived in the area and had historical knowledge of the impacts on the Blesbokspruit. Further, their interaction with water leads to their perceptions of what impact AMD STT has on the water quality of the Blesbokspruit.

5.3 Views on AMD in the Eastern Basin

> *Mining created it [the Blesbokspruit]. It became ecologically sensitive, very rich diversity of species and now mining is continuing to impact it.* (Pillay, personal interview, 2018)

This section presents the views on how AMD has impacted the eastern basin. These views include how AMD is understood, how severe the impact of AMD could be and how the Blesbokspruit has changed over the years. The time period, meaning when mining was predominant as opposed to when mining stopped in 2011, indicated what the reasons for the poor water quality were attributed to. These reasons stem from individuals' views and understanding of water quality.

For a century the East Rand was known to be a major gold-producing region with numerous high-profile mines and, therefore, increased acidic levels in the Blesbokspruit originated after these mining activities came to an end. This created the notion about the Blesbokspruit being formed by mine water decanting (Meyer, personal interview, 2018). Such notions are attributed to the way people perceive the characteristics of the area. From the literature on AMD discussed in Chap. 4, it is evident that there was a massive increase of water in the system from mining activity. Therefore, the public are aware that mine water contributes (water now flowing all-year-round) to the character of the East Rand. This is commonly understood by key individuals (i.e. community, government, researchers and activists) to be an unnatural condition (De Jager, personal interview, 2018b; Fourie, personal interview, 2019; Govender, personal interview, 2018; Meyer, personal interview, 2018).

Through water, the mining industry contributed to the economic growth of South Africa, and through mining, the flow of water of the Blesbokspruit has increased. This means that the volumes of water in the Blesbokspruit are co-produced through mining. "Mining created it, it became ecologically sensitive, very rich diversity of species and now mining is continuing to impact it", said Govender.[6] But the increasingly large volumes of water in the Blesbokspruit due to extensive mining contributed to the decline in the water quality. This is an indication of how social processes shape water and vice versa (Budds and Hinojosa 2012), meaning that social processes have constituted the waterscape of the area. With the large quantities of water now flowing, the pH profile of the Blesbokspruit was a strong indication of the declining water quality. Over the years the pH has been closer to 6 than 7,[7] resulting from AMD seeping (when pumping was sporadic during the 1990s) into the river (AMD lowers the pH) (Meyer, personal interview, 2018). A common understanding among residents, government officials, environmental consultants and activists is that the declining water quality from AMD was the main reason the Ramsar status of the Blesbokspruit was affected and the site placed on the Montreux Record in the 1990s (De Jager, personal interview, 2018b; Govender, personal interview, 2018; Madden, personal interview, 2018a; Meyer, personal interview, 2018; Pillay, personal interview, 2018; Van der Merwe, personal interview, 2018).

Owing to excessive mining, the extent of the water quality problem leading to AMD concerns was described based on how people viewed the government's ability to deal with the problem. Those interviewed either felt that the government did not

[6] B. Govender is employed by the DWS in the Chief Directorate: Mine Water Management Unit. He is at the forefront of policy development in mine water management.

[7] Water with a pH lower than 7 is regarded as acidic, and water with a pH greater than 7 is regarded as basic. A normal pH range in surface water systems is 6.5–8.5, and the pH range for groundwater systems is between 6 and 8.5. Drinking water must have a pH value of 6.5–8.5 to fall within environmentally accepted standards (Dirisu and Mafiana 2016).

understand the complexity of the problem (Meyer, personal interview, 2018) or understood the severity, but that the cost implications were too excessive and therefore government avoided seeking the required technical expertise (Bosman, personal interview, 2018a). The first concern arising from these views was with the way AMD was defined by the government to only include underground mining. However, key individuals felt acid rain was just as significant in defining the cause of AMD; for instance, water specialist Anthony Turton (cited in Kolver 2014) was of the view that the government had not yet grasped the complexity of the issue because neutralising AMD was not the best option. In his view, AMD does not only come from mine voids (the common understanding among government officials), but also when acid rain falls on tailings dams triggering the initial acidification process, leaching uranium concentrated in the crust (Kolver 2014; Naidoo 2017). Underestimating the impact of acid rain in an industrialised town such as Springs is problematic because many industries and open-cast coal mining contribute to pollution and pose an additional problem that exacerbates the impact of acid rain (van Zyl, personal interview, 2016). Acid rain is evident in rusty barb-wired fences, which can have an impact on the acidification of soils (Agricultural sector representative, personal interview, 2018).

Owing to how the government is known to refer to/define/understand AMD, it created perceptions about how it would be treated and whether the water quality would be improved. However, a government official from the Department of Water and Sanitation (DWS) noted that it was of importance to identify what the sources of AMD were in order to address it and acknowledged that it did not only come from underground mine workings. A large amount comes from waste facilities (tailings) situated on the surface (Govender, personal interview, 2018). This indicated the government's increased awareness of the sources of AMD. However, this awareness still does not include that of acid rain. The change in awareness can be explained through a time factor. When AMD concerns initially became popular in 2002 through the media, the government was still finding ways to approach the issue. Therefore, acknowledging the impact of acid rain had many implications, such as high costs and technical expertise. There was a tendency for the extent of the AMD problem to be described by various key individuals based on how water was understood and managed by the government. These individuals included various key individuals with different interests such as Springs community members, researchers, environmental consultants and the agricultural sector.

As a result, the way in which the government understood the problem amplified concerns around how serious the problem was. Vast ongoing criticism was publicised in the media (Creamer 2015; Kolver 2014; Springs Advertiser 2014b, 2015) prior to the implementation of the STT. Criticism centred on whether the STT would be implemented in time, due to limited funds, to avoid AMD breaching the environmental critical level (ECL) (rising and reaching surface water) (Kolver 2014), and because it was commonly known by the public that the western basin had breached (Meyer, personal interview, 2018). Further, the DWS could not manage the flow during the years of active mining, because the water became too excessive and therefore pumping large amounts rising from the mine void became increasingly difficult, leading to elevated sulphate levels (van Niekerk, personal interview, 2018). "Therefore, without the treatment, 106 million litres of AMD will be oozing out of the ground, causing infrastructural damage [and] the Blesbokspruit disappears; meaning three quarters of the Vaal catchment disappears. The impact is unimaginable when it's not attended to" (Pillay, personal interview, 2018). The government made positive strides by acknowledging that AMD was a serious threat and an emergency, according to environmental activist, Mariette Liefferink (cited in Kolver 2014), even though she had remaining concerns about the actual process for the STT. During the EIA process (Digby Wells Environmental 2014), a common understanding among stakeholders was that the project had to be implemented, meaning it was evident that the threat of AMD was serious. The limited information on the STT to which stakeholders had access influenced how they viewed and understood the extent of AMD.

The threat became evident in 2007/08 when the Gauteng Department of Rural Development (GDARD) was in the process of removing the Blesbokspruit from the Montreux Record. In 2009 they became aware of the fact that the Grootvlei Mine was releasing untreated water into the Blesbokspruit, presenting a noticeable difference in the water quality (Mashau, GDARD representative, personal interview, 2019). Further, once Aurora Empowerment Systems (owners of the Grootvlei Mine) stopped pumping in 2011, nothing was pumped until the treatment plant was in operation in August 2016. Therefore, nothing from the main mine void in the eastern basin should have been flowing into the Blesbokspruit because AMD did not breach at that stage (Holden, personal communication, 2019). Mines that could have been pumping and discharging water were liquidated, meaning they were no longer in operation. This means that if there were pollution occurrences, it was not from mining and it was not AMD seeping into the Blesbokspruit. If there were any form of AMD prevalent, then it was from unauthorised mining activity (Holden, personal interview, 2019). The GDARD representatives (personal interview, 2019) attested to this, suggesting that from 2010 to 2012 Grootvlei Mine stopped pumping and the water quality started to improve because nothing was coming into the stream. The acidic water was still below the ECL, meaning it did not breach, giving the government time to plan and implement the treatment. This implies that if the poor water quality then was attributed to AMD, it would have not been true.

The research conducted for this book found that the timeline was an extremely important factor in accounting for views and perceptions of stakeholders and key individuals about AMD, and its impact on the water quality of the Blesbokspruit. These perceptions are linked to how people construct their views of water quality to AMD, based on what they knew or heard. The AMD situation led to blame being assigned to the government for much of the problems with poor water quality due to lack of effective management to hold those mines accountable for the damages caused. However, what is not commonly understood by various individuals is that the poor water quality of the Blesbokspruit during that time (2011 until 2016) could have been from various sources other than what many attribute it to, namely, AMD. Other sectors that were discharging water into the Blesbokspruit could have been doing so without environmental authorisation and adherence to their water use licence. However, what the public were concerned about is that the different government departments did not have monitoring and compliance measures, and strict enforcement in place to identify and prevent these sources of pollution, which exacerbated the extent of AMD in their opinion. The research found that this led to conflicting views over who was responsible for the treatment.

5.4 Views on Who Is Responsible for AMD Treatment

> *Mines that were operating during the booming years of mining, did so 'quite legally' because environmental laws and 'clean' operating practices were not developed fully.* (Meyer, personal interview, 2018)

When AMD reached crisis proportions in the Witwatersrand goldfields in 2002, the main debates that surfaced were around who should take responsibility for cleaning up this environmental burden (Ambani 2013, p. 88; Naidoo 2017, p. 53). These debates are ongoing due to the excessive cost attached to AMD treatment, and they remain deeply rooted in political issues. This section presents the conflicting views among individuals on who is responsible for both the STT and long-term treatment (LTT) of AMD. Debates on who should take responsibility for AMD treatment form an important part of how individuals socially construct the water quality of the Blesbokspruit. This indicated that perceptions were due to the highly political nature of the problem when implementing the short-term solution, which reaffirmed the fact that water and mineral governance in South Africa are extremely politicised (Mpofu et al. 2018, p. 78).

The first concern was that the mines that were responsible were no longer in existence, and this led to major environmental problems because policy and water governance did not take into account the sustainability of the water resources. The Mine Water Management Policy (DWS 2017, p. 2) attests that "while the challenge [AMD] is limited to the mining sector during operations, it eventually becomes externalised during mining downturn, and is especially pertinent post-mining closure, especially if mine closure does not proceed according to regulatory approved recommendations". Financial provisions made by mines were not sufficient to deal with the present mine water impacts (DWS 2017, p. 8). In the view of van Niekerk (personal interview, 2018), the public felt that in the past mines were not held liable, more so because when these mines were liquidated they did not have to cover the costs and public money (through tax) was used. "Now-abandoned mines that were operating during the booming years of mining, did so 'quite legally' because environmental laws and 'clean' operating practices were not developed fully" (Meyer, personal interview, 2018). Therefore, while mining was contributing to the economic growth of South Africa, the government did not have proper environmental management structures in place.

The DWS is the authoritative national government department for water and has to manage degradation of South Africa's water systems. However, the confusion over which government department is responsible stems from the fact that mining rights fall under the jurisdiction of the Department of Mineral Resources (DMR). The lack of clarity stems from how the DWS was forced to control further environmental degradation as the custodians of the country's water resources. One of the aims of the Mine Water Management Policy is to rectify and clarify positions of authority between spheres of government, and furthermore, to align the roles and agreement between the DWS, the DMR and the DEA, who all form part of the regional mine closure plan process (DWS 2017, p. 5). This was an issue of policy processes not being explicit, and because this was unresolved, the LTT will be implemented much later than intended due to cost implications. In the interim, increased negative impacts on the Vaal River will occur due to excessive salinity from the discharged water brought about by the STT (Liefferink, personal interview, 2016). This means that the DWS will have to control this salinity level until governmental responsibilities and management processes, related to who is accountable, and the issue of funding availability, is clarified. Lack of clarity has led to varying individual views on who should take responsibility, which has created negative perceptions of the STT.

Stakeholders from local government questioned what the cost per litre was to treat AMD water in the short term, and who would remain accountable for these costs (Klip River Forum 2018). Tlale of the Trans-Caledon Tunnel Authority (TCTA) (Klip River Forum 2018) reported that the exact amount was unknown, but about R10 million per month was spent mainly on electricity costs. The DWS is covering the costs for the STT and has partnered with some mines. However, as regards the LTT, "who will pay is unknown" (Tlale in Klip River Forum 2018). To clean up AMD in the short term the estimated capital expenditure is R721 million, and in the long term, R2.97 billion (Bega 2021). According to Liefferink (personal interview, 2016), the Water and Sanitation Minister announced that because such large amounts of money were needed, it was the public's (taxpayers') money that was covering these costs, and, in future, they would continue to contribute to funding the LTT through increased tariffs for water. Some of the cost would be recovered from the remaining gold-mining companies (four or five that are operational on the Witwatersrand) and will be contested by the DWS through the Chamber of Mines (Liefferink, personal interview, 2016). The assumption is that these costs will not be recovered and the public will carry them in the end, which means that poor water quality (for the damages caused by the mining industry who profited) will actually be addressed by the public and new mining companies (Liefferink, personal interview, 2016).

Liefferink (personal interview, 2016) felt that current and new mines should not be held accountable for over 100 years of mining, when historic mines are not paying the price for their environmental damage. This is where the public frustration over who is responsible for the clean-up developed (Maurizi, personal interview, 2018). She stated that it was not possible to hold historic mines accountable for this problem as they could not come back and clean it up (Maurizi, personal interview, 2018). According to Fourie (personal interview, 2019), this is unacceptable because it boils down to what is the ethically correct decision to make. Even though these abandoned mines were responsible, taking ownership of a mine comes with environmental responsibilities as well, explained Maurizi. The responsibility was not taken and is partly due to the government's lack of management and lack of political will to hold mines accountable. "Who was the authority that allowed this [environmental damage]? There were supposed to be funds available for the clean-up" (Fourie, personal interview, 2019). The government has a policy obligation to ensure that mines were operating according to their water use licence and mining right (Fourie, personal interview, 2019). The confusion over who should be held responsible resulted from stakeholders who have their own interests, such as employment needs (public) or economic growth (the DMR and the DWS), and this is why environmental protection was not adhered to (Maurizi, personal interview, 2018). Therefore, there was no point in issuing a mining right if no compliance and enforcement measures were in place to ensure that the user was meeting the terms of that mining right (Maurizi, personal interview, 2018). This can be illustrated by the court cases with former directors of Aurora Empowerment Systems who were accused of causing environmental damage when taking ownership of the Grootvlei Mine. While the directors were found guilty in 2015 and liable for R1.7 billion in damages, the case was dismissed without charges pursued (Sicetsha 2020).

Stakeholders such as local government, researchers, community members and environmental activists felt that it was the national government's responsibility to clean up the environment (Klip River Forum 2018) because the government's interests in economic growth exceeded that of environmental protection, which led to a lack of environmental (and also water) management and political will to enforce compliance. The lack of management and poor governance meant that there was no transparency in holding mines accountable, and the public do not know what happened to the rehabilitation funds to which historic mines had to contribute (Fourie, personal interview, 2019). This created a barrier in trust between the national government and the public. Tlale (personal interview, 2018a) understood that the negativity had more to do with the accountability of those who were responsible for the damages caused and whether they were held accountable. Tlale, although employed by the DWS, indicated that the national government was aware of how the public felt. This links to trust issues from the community, activists and local government, with national government to hold those mines accountable, linking it to an issue of poor governance.

Ultimately, the state has the power and the influence to make better governance choices, however unpopular they might be (Meyer, personal interview, 2018). The crux of the issue and the major concern are that the public has to pay for the environmental damages caused by those mines who have made their millions and have now gone away (Fourie, personal interview, 2019). Stakeholders from local government, researchers, environmental activists and community members find it unfair for the public to pay for something from which others had made a lot of money. The clean-up costs seem to be the public's responsibility through the power of the national government to fix this environmental problem. The research found that as a result, negative perceptions about the STT developed because the costs remain so high and the water cannot be used for drinking purposes in an increasingly water-stressed country. The government should take responsibility to clean up the mess, according to many individuals, which contributed to initial perceptions about the implementation of the STT.

5.5 Views on AMD Short-Term Treatment Before Its Implementation

I wonder how this water, treated or not, will change the face of our beloved Blesbokspruit. (*Spring Advertiser* 2014d)

Why do you have to dispose the sludge here; why don't you take it and put it in your backyard. (Springs community member in Digby Wells Environmental 2014)

This section presents the views of interviewees on the STT treatment before it was implemented. The relevance of this section is to indicate how the perceptions of the STT were formed because an EIA was not conducted for the AMD treatment plant and discharge of neutralised water or for the sludge disposal site in the mine void. Therefore it is relevant to present the concerns that were raised during the EIA public participation process for the above-ground Grootvlei Site 6/L/16 TSF. Even though this site was not used when the STT was implemented, it is important to refer to the concerns raised during this time in order to understand why and how perceptions of the STT developed, and how these perceptions influenced various social constructions of water quality of the Blesbokspruit (Digby Wells Environmental 2014; Digby Wells Environmental 2015, p. 7).

In 2014, various stakeholders, referred to as interested and affected parties (IAPs), attended meetings for the scoping phase of the EIA for the selection of a sludge disposal site as part of the STT of AMD. IAPs included those who attended the Blesbokspruit Forum and the general public. Stakeholders were of the view that their comments and opinions could influence the outcome based on their need and use of the water. Further, the purpose of an EIA is to hear the public's views and identify possible impacts on the community and the environment. These views are presented below and are taken from research conducted by Digby Wells Environmental (2014) entitled *Environmental impact assessment for the construction of the proposed sludge disposal facility and pipeline associated with the treatment of Acid Mine Drainage from the Eastern Basin of the Witwatersrand, Gauteng (Full Comment and Response Report).* During the EIA public participation meetings, IAPs had a chance to raise their concerns about the sludge disposal site, Grootvlei 6/L/16, with a respondent from Digby Wells providing answers to or recording questions and comments. While it was explained that the significance of implementing the STT was to prevent raw AMD decanting into the Blesbokspruit, IAPs still raised concerns. These concerns were based on perceptions of how government would manage the water quality, keeping in mind that disputes still existed over who should take responsibility for the costs of treating the water.

After reviewing the EIA comments, it was found that IAPs were concerned with the use of Grootvlei Site 6/L/16 because of the possible impact on the Blesbokspruit for various reasons, including their property values, assets and employment (Anticevich 2014; De Jager, personal interview, 2018b). Many of the questions raised by IAPs in the public participation meetings were not only linked to the role of Digby Wells, which was to conduct an EIA only for the site where the sludge would be disposed. IAPs used this meeting to raise their concerns about the impact of the neutralised water on business, agriculture, tourism and property values.

Below are some of the concerns that were raised during the public participation meetings. Note that some of the comments raised during these meetings referred to the discharge of water rather than the site for the disposal of sludge for which the meeting took place. The reason for this was that there was no platform for IAPs to raise concerns about the volume and high sulphate levels of the discharged water. Further, these concerns stemmed from the possible impacts on their use of the water and its quality of the Blesbokspruit. The concerns were as follows:

- Water pollution can affect the fish in the Blesbokspruit.
- Loss of biodiversity due to water contamination.
- Air pollution and the smell from the treatment plant.
- Possible groundwater contamination, which is used as drinking water for people and animals in the area.
- Lack of credible research was prepared on boreholes, the report incorrectly recorded the number of boreholes in the area.
- Birdlife will be affected due to water quantity; additional water has already altered the habitat, causing significant reduction in waterfowl. Disposal of neutralised water into the wetland will worsen the situation.
- Farmers, particularly subsistence farmers, are ill-equipped and poorly informed about how to handle the levels of salinity.
- Possible financial loss for farmers and job loss resulting from irrigation with saline water and no salinity control (will be a significant cost to the farmer).
- The STT is meant to remove the metals from the acidic water. However, the proposed sludge disposal site in close proximity to the Blesbokspruit will re-introduce metals into the system.
- The pipe could leak and the sludge could end up in the Blesbokspruit and pollute the wetland.
- Who will take responsibility for the sludge disposal site after it has been decommissioned since the STT is meant to continue for 5–8 years.
- The neutralised water (high sulphates) will add more contaminants to the river.
- Increased sulphate levels in the water can impact on environmental health, agriculture and human quality.
- The sludge can be radioactive; therefore, to ensure radioactivity on the existing sludge, the disposal site does not increase.

Based on these concerns, IAPs were deeply concerned about the STT of AMD. At this stage, several IAPs suggested that the mine void be used for the sludge disposal, based on the concerns indicated above, but were informed that it would impact on groundwater (see Sect. 5.6.1). With the insistence on the use of Grootvlei Site 6/L/16 and the discharge of neutralised water into Blesbokspruit, GDARD (in Digby Wells Environmental 2014) said the STT would be "economically impossible," because of the potential human and ecological harm downstream of the Blesbokspruit. Due to the Blesbokspruit being placed on the Montreux Record, concerns around the significance of the Blesbokspruit were raised (De Jager, cited in

Digby Wells Environmental 2014). IAPs felt that the disposal of sludge at Grootvlei Site 6/L/16 would compromise the water quality even further, and therefore it would be even more difficult to have the spruit removed from this record. The CoE Water Quality Manager, Maurizi, raised her concerns about the management of the STT because the treatment could create a more expanded river in the future, with the plant situated in close proximity. An expanded river will change the landscape of the area (Pillay, personal interview, 2018). Birdlife South Africa was concerned that "it's one of the Vaal River's larger tributaries and its catchment covers 1 000 km^2, ultimately supplying water to the densely populated Gauteng Province" (Digby Wells Environmental 2014). The wetland already assists as a purifier of industrial effluent, and this has altered its flow. The STT will add more water to the system and might over-exacerbate it, impacting on the diversity of birds, many of which are water birds that are found in Marievale Bird Sanctuary, which contributes to the attractiveness of the East Rand (Digby Wells Environmental 2014). Besides the high salt content that the STT would add to the Vaal River system, other factors negatively impact on the already sensitive environment and nearby residents. These impacts include existing industrial effluent discharge; abandoned gold-mining operations; historical coal mining; complex dolomitic geology (including sinkholes situated opposite the proposed sludge facility); AMD; wastewater treatment works discharges; the CoE wastewater pump stations that fail and become flooded regularly; untreated sewage and runoff from informal settlements; illegal mining operations; illegal dumping; flooding and reed encroachment of the river (*Spring Advertiser* 2014a).

Owing to the array of existing problems, the Federation for a Sustainable Environment (FSE) was sceptical of the STT and the exemption of an EIA for the construction of the treatment plant added to these concerns (Digby Wells Environmental 2014). This led the FSE to express little hope that their comments on the EIA would be incorporated into the decision-making process. The interactions that various IAPs from the community, local government and provincial government, agriculture and tourism industry have with the water resulted in concerns related to the management of the STT. As a result, they asked to be included in the process by receiving feedback to reassure them that the process would run accordingly. Suggestions included that potential impacts and monitoring results of the STT be disclosed; the community be involved in monitoring the water; and that basic mitigation measures be put in place for the sludge disposal facility. A detailed hydrocensus was also recommended for surrounding water users to manage potential impacts of groundwater. Digby Wells indicated to the IAPs that results and monitoring plans were to be submitted to the DEA, the Blesbokspruit Forum and the Blesbokspruit Trust, and could be accessed through these platforms. IAPs were informed that the costs of solutions to water users were minimised, and that continued monitoring of quality trends and review of whether reduced pumping might contribute to improving water quality (Digby Wells Environmental 2014). The importance of raising these points is to indicate how the process was communicated to the public and what access they had to information on the STT.

Despite the concerns raised above, Digby Wells received support from IAPs for the project as they showed awareness that the treatment had to be implemented. Birdlife South Africa believed many positives would come from the project if correctly implemented. The CoE had no objection to the proposed facility and pipeline routes, as without implementing the STT there would be uncontrolled decanting of AMD into the environment. This could result in severe ecological and social impacts eventually reaching the Vaal River, and added environmental, health and agricultural impacts and on the quality of human life if not addressed. The area has the most beautiful agricultural land and best soil, and therefore the treatment was necessary (De Jager, personal interview, 2018b). For these reasons, approving Grootvlei Site 6/L/16 for the sludge disposal would prevent pollution of surface and groundwater and surrounding environments (Digby Wells Environmental 2014).

To prevent such impacts, the legal requirement for an EIA was evident. IAPs' various interests in water, indicated by their concerns raised, showed that the way in which the water was managed played a pivotal role in people's perceptions and acceptance of the STT. IAPs knew the STT was necessary and wanted to be involved in the process through inclusion in water management, because the process directly impacted on their various interests. This led them to request that water governance be approached holistically. The next section explains the perceptions after the implementation of the STT.

5.6 Views on AMD Short-Term Treatment After Its Implementation

> *AMD treatment has a bad reputation out there and I think it is going to take a lot of effort to change the mind-set of the people.* (Fourie, personal interview, 2019)

> *This project is meant to make our lives better.* (Tlale, personal interview, 2018a)

The STT process started in August 2016, and the AMD treatment plant was launched in February 2017. Based on the EIA process (discussed in Sect. 5.5), the community developed perceptions about what the location of the treatment plant would have on their property values, considering that there were many residential areas surrounding the AMD treatment plant's location. Even though the Grootvlei Mine's Shaft No. 3 met all the requirements for the treatment plant to be constructed at this site, there were concerns from the community about the location of the plant and the possible noise and smell. However, though monitoring the plant does not cause much dust, noise is minimal and there is a slight smell of sulphur in the water when it comes in contact with air, but this is not frequent (Tlale, personal communication and site visit, 2018b). Tlale suggested that there was an additional safety component with where the plant is situated, making it much safer for communities because the site was previously filled with Zama Zamas[8] and illegal mining (Tlale, personal communication and site visit, 2018b). Tlale felt that the community should be proud of the treatment plant since it is "the biggest in the world," of good standard and it was clean, which was a massive achievement, which should be an honour for those living in Springs.

Key individuals from provincial and local government, community members, and the tourism and agricultural industries are of the same positive opinion about the STT. "It is one of the best treatment processes" (Mashau, GDARD representative, personal interview, 2019). "I am not a water specialist, but obviously reverse osmosis (desalination) will be the best, but it is very expensive. What they [the DWS] are doing now is probably a cheaper alternative because they can't afford reverse osmosis at present" (Agricultural sector representative, personal interview, 2018). "It is true, the water is better in quality than before the treatment" (Van der Merwe, personal interview, 2018). "It's top class" (Naidoo, personal interview, 2018). These quotes indicated that key individuals were impressed with the general efficiency and standards when taken on a tour of the plant (Humphreys, personal interview, 2018; Madden, personal interview, 2018a; Mashau, GDARD representative, personal interview, 2019; Naidoo, personal interview, 2018). Madden indicated that the DWS endured a lot of pressure to get to where they are with the STT; "I am prepared to back them in this regard" (Madden, personal interview, 2018a). However, despite the positive comments and feedback, and the fact that the treatment plant is the biggest in the world, that was not enough. The interest of the various stakeholders was with potential impacts regarding the quality of the water and the plant was not yet producing desalinated water because the LTT is yet to come.

When the STT began, stakeholders (all IAPs) were under the impression that the above-ground sludge disposal site (Grootvlei Site 6/L/16 TSF), for which an EIA was conducted, formed part of the actual STT process. Further, they expected their views and concerns from the EIA public participation process to be taken into account, and that they would be informed of the monitoring and management of the STT. Stakeholders who were involved in the Blesbokspruit Forum questioned national government decisions and the significance of an EIA process because they felt misled. Reasons for these decisions remained unclear, mixed perceptions of the STT developed and varied among stakeholders, creating negative perceptions of the sludge disposal and the perceived impact of discharged water.

The purpose of the following three subsections is to present the perceptions that suggest the complex relationship between socio-political and historical contexts, coupled with the different responsibilities and implementation strategies of the STT by the DWS. This resulted in tensions between the national government (the DWS) and other stakeholders, such as the public and other departments responsible for the management of the Blesbokspruit, implying the existence of power relations (discussed in Chap. 7).

---

[8] Term used for illegal miners.

The first subsection explains what happened when the public became aware of the fact that the sludge disposal site had changed to the mine void without their knowledge and without an EIA being conducted, and what their perceptions of the use of the mine void were. The second subsection sets out perceptions about the discharge of neutralised water, also exempted from an EIA. The third subsection discusses the perceptions of the significance of an EIA process due to people's time being invested in this process for the Grootvlei Site 6/L/16, which was not used once the STT was implemented.

5.6.1 *Perceptions of the Underground Disposal in the Mine Void*

> *To stand at the top and pour it back in, does not make sense.* (Barker, personal interview, 2018)

During the EIA process, stakeholders (such as IAPs), including community members and local government officials, suggested that the mine void be used as the sludge disposal site. They were informed by the environmental consultants that it was not an option, as it could impact on groundwater resources (Digby Wells Environmental 2014). Therefore, the use of the mine void was regarded as flawed due to possible groundwater contamination. In Pillay's view (personal interview, 2018), environmental consultant at Digby Wells, using the mine void was initially seen as a flaw; disposing of the sludge down into the mine void for a plant that operated at up to 110 mℓ/day[9] was a mistake. This suggested that negative views about the use of the mine void were most possibly formed based on the EIA public participation process. This resulted in perceptions that the mine void was flawed as a sludge disposal site. These perceptions resulted from the public being "misinformed" by respondents (environmental consultants) during the EIA public participation process (Maurizi, personal interview, 2018; Tlale, personal interview, 2018a), which created trust issues between national government (the DWS) who implemented the treatment and other stakeholders who were unaware of this decision. In reality, the mine void is what the public initially suggested for the sludge disposal, instead of the above-ground TSF, but because they were informed that it could impact on their groundwater sources, it led to changes in their perceptions. This situation indicated that there was a deep political nature in the solution to AMD, linked to lack of communication and possible attribution to a power play.

During the interviews conducted, key individuals expressed their deep concerns about the use of the mine void. These concerns mostly surfaced because they were either not aware that the sludge was disposed of in the mine void for the STT or why the site had changed without them being informed. Various perceptions were formed

[9] The sludge is disposed at about one-and-a-half to two kilometres deep, and abstraction happens at about 5–6 km deep at Gold Mine's Shaft No. 3 (Pillay, personal interview, 2018).

based on this lack of information. They include whether there was enough space in the void for the sludge disposal to avoid it contaminating the partially treated water (Madden, personal interview, 2018a), because it is not known what is happening underground (Pillay, personal interview, 2018). Madden illustrated the severity of his concerns:

> *I am very uneasy about the disposal of sludge, where it is being disposed now and the future disposal. I have had 33 years of experience with the mines on the East Rand, and that qualifies me to make a statement that I am unhappy about that [the sludge] going down, because how long will the void be able to take it? Nobody knows. Nobody goes down there anymore, and we are playing with a catastrophe and they [TCTA] admitted it to me when I went to the plant.* (Madden, personal interview, 2018a)

This indicated that key individuals felt that not enough knowledge and information had been shared to know what the impact of the sludge being placed in the mine void was (Fourie, personal interview, 2019). Further concerns raised due to the lack of awareness included that using the mine void could restrict access to possible mineral resources. "We have created a hole which is access to a resource and by filling that hole with sludge we are limiting potential to re-access that resource" (Barker, personal interview, 2018). Even though most of the tailings on the East Rand have been reclaimed, there is still gold left (in the mine void) where the sludge is being deposited (Barker, personal interview, 2018; Fourie, personal interview, 2019). "To stand at the top and pour it back in does not make sense, because what if obstructions are caused" (Barker, personal interview, 2018). Barker, a consultant at Ergo Mining, has in-depth knowledge of reclamation, and this led him to believe that extracting the mineral resources left in the mine void could be profitable because technology continually improves. The development in years to come might be more phenomenal, and using the mine void could restrict access to the resource (Barker, personal interview, 2018). Barker is a predominant voice in the Blesbokspruit Forum, often reiterating the need for government to take action. However, he also points out the value of mineral resources. These perceptions stemmed from the fact that there was a lack of awareness, and this comes from a lack of communication.

Consequently, the above perceptions stemmed from the fact that the use of the mine void was exempted from an EIA and, therefore, no water use licence for this activity exists because the use of the mine void was labelled a 'pilot project' (Tlale, personal interview, 2018a), meaning the risks are unknown. Bosman (personal interview, 2018a), based on her expert knowledge on environmental and water-related processes and policies, suggested that there has to be a water use licence for underground disposal because the licence explains the pros and cons of the disposal. The disposal process involves "disposal of waste or water containing waste in a manner that may detrimentally impact the groundwater resource" (Bosman, personal interview, 2018a; Bosman 2018b). The licence is meant to be publicly available in terms of the *Promotion of Administrative Justice Act 3 of 2000* (PAJA) (discussed further in Chap. 7).

An EIA is a lengthy and costly process, and cannot be conducted for no reason (Tlale, personal interview, 2018a). Only after the TCTA were ready to pump and treat the water did they realise that they had a conflict with the chosen site for the disposal of sludge. The treatment was already scheduled to begin, and they had to start disposing of the sludge because the main priority for implementing the STT was to protect the ECL, and the financial implications had to also be considered. Other options were explored, including conducting another EIA to dispose of the sludge at an alternative mining site, which will be done for the LTT. However, the lengthy nature of the EIA process made this difficult. However, through discussions with the AMD project team, the option of putting the sludge underground (in the mine void) surfaced. The use of the mine void became an option and deliberations between the DWS, the DMR and the Department of Environmental Affairs (DEA) about whether it would work took place (Govender, personal interview, 2018; Tlale, personal interview, 2018a). In the past, the option of using the mine void was never explored to the satisfaction of everyone, and it was therefore not previously seen as a viable option (Tlale, personal communication and site visit, 2018b).

The DWS recognised the potential impact of using the void on the water resource (Blesbokspruit). Tlale (personal communication and site visit, 2018b) acknowledged that there was a risk because so much was unknown, and the possible risks (if any) can only be determined or known once the project is complete. With that in mind, the sludge disposal was subject to strict criteria being met (Govender, personal interview, 2018). Monitoring of the quality of the water is conducted on a monthly basis to see what the impacts of the sludge disposal are. The biggest challenge was whether the water quality would be compromised as the sludge was disposed of. The concern was that if it rises, it can affect the pumps, which are an expensive part of this process and project, and could not be risked (Tlale, personal interview, 2018a).

The fact that this was a pilot project, and that the DWS and the TCTA did not know the risks themselves, meant that the parties involved could not communicate the site openly to the public. This could be due to fear of how the public would react and because they could not answer many of the questions the public would pose. "At any point anyone can go to the TCTA or the environmental consultants and say that they don't know enough about what's happening underground and they will be right. No one knows enough about what's happening underground" (Pillay, personal interview, 2018). Using the mine void is more effective from a cost perspective. The only way to understand this further is to drill holes, and that is expensive (Pillay, personal interview, 2018). Underground disposal can be feasible. However, the management measures can only be determined after underground hydrological modelling has been conducted to understand the flows, volumes, qualities, dispersions and so on, and then only can decisions be made on if it is manageable (Pillay, personal interview, 2018). If the mine void works, then it would be the ultimate solution to the treatment and can be used in future (Tlale, personal interview, 2018a). Water specialist Anthony Turton (personal communication, 2018) finds it to be good practice to remove surface waste and dispose of it back into the mine void where it came from. This practice restricts the places that AMD can form, it has less effect on the environment and the sludge is deposited where there is a higher pH, whereas AMD has a lower pH, therefore it "tweaks the chemistry" of mine water (Govender, personal interview, 2018). By actively managing the ECL and ensuring that it is kept below the lowest point of the aquifer contact zone, no matter how bad the void water is, it can never contaminate groundwater floating in the aquifers above (Turton, personal communication, 2018). It also presents a way to open up land. Filling these voids instead of highly valuable ground that would otherwise be used to dispose of sludge can create new opportunities for development (Turton, personal communication, 2018). It is therefore a positive step that reduces visual and human health-related impacts (Fourie, personal interview, 2019).

Bosman (personal interview, 2018a), an expert in water use licence applications, suggested that someone who did not understand what was happening would think of how the sludge could impact on the water and that was why they reached their own conclusions. Geologists are of the view that it is not uncommon to dispose of sludge in a mine void, whereas community members who have limited knowledge are of the opinion that it is harmful, especially when the implementing agents have no insight into what social and environmental risks can occur (Bosman, personal interview, 2018a). "If they [the TCTA] can provide more certainty on the impact of the sludge disposal into the mine void, then stakeholders will be more at ease and supportive" (Madden, personal interview, 2018a). For these reasons, it was suggested that experts should be included to provide advice on what should be done (Bosman, personal interview, 2018a; Pillay, personal interview, 2018). Those who understand the technical nature need to be involved in the process and look for the gaps/unknowns (Bosman, personal interview, 2018a).

The information available is that the sludge is disposed of very far from the pollution level, and even if it had to rise, there are indicators to identify an increase in pollution. The TCTA is working far below the level that can lead to environmental impacts required to ensure that it does not breach (Tlale, personal communication and site visit, 2018b). This means that extra precautions are taken, and should a problem arise, there is room to address it. The TCTA's main concern is ensuring that the ECL is maintained. As a result, there is a level at which the sludge is kept, because with water the lower it is discharged, the more expensive it becomes. The water is pumped to maintain the level, and this process can continue for as long as it is needed and will not create an impact on the water quality (Govender, personal interview, 2018; Tlale, personal interview, 2018a). The DWS is working on smaller projects in terms of the rising water in the underground eastern basin mine void, which is currently being maintained at a level where it will not impact on the surface resources or on the groundwater sources (Govender, personal interview, 2018). The entire eastern basin has large amounts of dolomitic reserves with groundwater and surface water resources that are quite close to the surface, with the result that it becomes impacted. The reversal of impacts can take a long time. Govender (personal interview, 2018) believes that the water is safe for now. Given that the STT is in place, there is no impact or the impact, if any, associated with underground mine water is minimal.

How would one know if something works, if it has not been tried and tested, and then, ultimately, how do processes improve in the future if various options are not considered? Unfortunately, one generation could suffer the consequences of a pilot project to make processes better for another generation, which definitely impacts on the aims of sustainable development. The DWS and the TCTA were simply in a position to make a decision; they were in an authoritative position to act. However, stakeholders were already negatively inclined towards the STT because they felt misled, misinformed from the EIA process and excluded through the lack of communication, and therefore various perceptions of the STT were formed based on these unknowns. Through fear of the questions that could be raised and because of the unknown factors, the DWS did not openly communicate the new sludge disposal site to the public. This implied that trust issues existed between the national government and the public.

More important to consider is that the issues of trust influenced the relationship between stakeholders. Stakeholders' concerns about the use of the mine void stemmed from a lack of communication. "TCTA and DWS haven't done enough to create awareness for the community, so that they can appreciate the development and progress made, this is something that we need to do" (Tlale, personal interview, 2018a). Tlale suggested that taking members of the Blesbokspruit Forum to explain how the plant operated would be useful. However, this meant that even the people involved in the management of the Blesbokspruit were unaware of the mine void as the sludge disposal site. Further, if the mine void was initially considered flawed, how can stakeholders trust that the monitoring is sufficient and that it would not impact on their groundwater? The lack of awareness led to many negative opinions and perceptions being formed on the STT process, and this is due to the interest of individuals regarding their use of the water.

5.6.2 *Perceived Impact of Discharged Water on the Blesbokspruit*

> *Nature has its own balance, and it's being messed with quite badly. The AMD treatment plant is meant to rectify some of these imbalances, but I am not convinced of that.* (Humphreys, personal interview, 2018)

In addition to the negative perceptions on the disposal of sludge into the mine void are negative perceptions of the discharge of the partially treated (neutralised water) AMD water into the Blesbokspruit. Two concerns stem from the water that is discharged into the Blesbokspruit. The first involves the volumes of water altering the flow, and the second is the quality of the water due to the high sulphate content remaining in the discharged water that could reach and impact on the Vaal River system. As explained in Chap. 4, the short-term intervention involves partially treating the water, meaning it is not potable when released. This section presents the perceptions of stakeholders on the discharged water, which influence the social constructions of the water quality based on the flow of the water and the technology used (high-density sludge) to treat AMD. The aim is to present how water is framed socially, and perceptions of the STT are based on the interests of stakeholders and their need and use for water.

As already explained throughout this chapter, perceptions stem from the fact that no EIA was conducted for the treatment plant, including the discharge of neutralised water or the disposal of sludge. This means that a water use licence has not been issued for the STT because no environmental and social impact assessments exist to determine the water use requirements (Bosman, personal interview, 2018a). This meant that the impact of neutralised water has potential consequences that are unknown. Individuals' perceptions, based on these unknown consequences, qualified them to question the legitimacy of the STT and concerns surrounding when the LTT (desalination/potable water) will commence (Liefferink, personal interview, 2016). Maurizi (personal interview, 2018), a resident and employee of the CoE, was concerned about the amount of time that had passed; "the process [STT] was delayed and nothing happened for a long time. Full desalination should have taken place years ago, and now they are only doing half the treatment. Is desalination really going to happen at all?" "The government dragged their feet, and therefore the circumstances called for immediate action" (Van der Merwe, personal interview, 2018). These statements indicate the lack of support from the public, and Tlale (personal interview, 2018a) was of the opinion that it was because the water is not being treated to potable standards as yet, resulting in negative media reports continuing to surface, in order to push the government to implement the LTT. The public feel misled, and the AMD project team was aware of this but felt the STT was a positive step forward.

Tlale of the TCTA and Govender of the DWS, AMD project team members, explained the significance of the difference in the water quality as being safer for the environment compared to before it was treated. The neutralised water is only part of the STT, and not the LTT (Govender, personal interview, 2018; Pillay, personal interview, 2018; Tlale, personal interview, 2018a). The effects of raw AMD are beyond imagination, whereas with the neutralised water, there are effects that cannot be denied, but as compared to the effects of raw AMD, it is significantly less (Pillay, personal interview, 2018). It was not merely an issue of the surface water being polluted. If the water was left untreated, it would rise and there could be additional impacts (Govender, personal interview, 2018). Such impacts could be seen on underground water too, and this will be worse for the environment and the people who use groundwater (Tlale, personal interview, 2018a).

The STT uses alkaline material to bring the pH to a suitable level and in the reaction metals are removed, recovering about 40–50% of the salinity, which is the sulphate irons (Govender, personal interview, 2018). However, 50% of salts are still going into the Blesbokspruit. The impact of this remaining salinity in the Vaal River system is meant to be managed by the DWS to prevent compromising downstream water users (*Spring Advertiser* 2014c). The water has a sulphate level of 2000 mg/ℓ as opposed to 4000 mg/ℓ when left untreated (Govender, personal interview, 2018). Therefore, Govender does not foresee a challenge with the neutralised water because at 2000 mg/ℓ as it flows further downstream, the sulphate drops to about 800 mg/ℓ because there is a dilution effect. Therefore, it is still safe for users such as agriculture. "Does this have an impact or is it ideal? No it's not ideal, it has an impact to a degree, but that impact will be localised to the immediate stretch of the Blesbokspruit, because there is dilution effect. It's not the quality that is acceptable to the environment even though it is safer but diluting the salt (STT) mitigates the pollution of the water" (Govender, personal interview, 2018). The dilution effect lessens sewage pollution (Fourie, personal interview, 2019; Govender, personal interview, 2018; Tlale, personal communication and site visit, 2018b). Sewage pollution is problematic due to dysfunctional wastewater treatment plants (Fourie, personal interview, 2019). *E. coli* is evident in the water and was often reported at unacceptable levels in the Blesbokspruit Forum (Blesbokspruit Forum 2018a, b, c, d, 2019a, b, c, d). Tlale (personal interview, 2018a) explained that the impact *E. coli* has on the water is a result of the municipality not treating the water adequately, which can have health impacts, which are worse than AMD. She is suggesting that the dilution from the discharge is beneficial.

Despite this dilution effect improving the quality of the water, Bosman (personal interview, 2018a) suggested that determining the state of the water was based on what was sampled. If a sample of fresh runoff is taken, with the partially treated water, then it will appear as if it is better in quality. However, if the total load of salts discharged into the Blesbokspruit are examined, it is evident that heavy metals do not dissolve in the water, and with the potential that it is exposed, it will seep into the sediments, leading to long-term risk and long-term release (Bosman, personal interview, 2018a). Liefferink (Blesbokspruit Forum 2018a) raised concerns about the different sulphate levels indicated in Rand Water's and the DWS's water quality samples and was also concerned about what impact the choice in sample point had on representing the true quality of the water, as Bosman indicated above. Liefferink (Blesbokspruit Forum 2018a) suggested that if samples were taken closer to the dilution points, the high levels of sulphate would cause or avoid (depending on the sample point and dilution effect) further issues downstream where Karan Beef Feedlot, for instance, is situated. In 2018, Rand Water (in Blesbokspruit Forum 2018c) reported that for the second quarter of the year, the discharge of the neutralised water for the first time reflected non-compliant sulphate levels, but it did not exceed 1000 mg/ℓ, which reflected immediately downstream of the discharge point. Tlale (in Blesbokspruit Forum 2018c) clarified that high sulphate levels were recorded downstream but were not connected to the STT operations and discharge of the neutralised water, based on the water samples conducted by the TCTA. This indicated that other sources of pollution contributed to the quality of the water due to the known discharge of neutralised water. The positive versus the negative impacts of the discharged water are measured based on complaints received and implied that the TCTA and the DWS were taking note of the concerns (Tlale, personal interview, 2018a).

Tlale (Blesbokspruit Forum 2018a) reported that the water quality of the eastern basin was better than the other basins of the Witwatersrand and was within the directive limits for the STT. The increased sulphate problem has to be further investigated, but the plant was running better than expected and the Blesbokspruit would be far worse off without it (DWS representative in Blesbokspruit Forum 2018c). Another sustainable technology was assessed, but despite the delays, the STT is continuing. The LTT is meant to resolve the high salinity impact on the Vaal River system, but the STT will be implemented for a period of 5–8 years before this happens (Digby Wells Environmental 2015, p. 3). Members of the AMD project team acknowledged that the sulphate level was not ideal, but described the water as in a better state, which is also a way of trying to increase the public's support for the STT (Tlale, personal interview, 2018a; Govender, personal interview, 2018).

The DWS tried to create awareness among communities and other stakeholders to justify that this option was better than no option at all. The DWS tried to gain acceptance of the STT from stakeholders by explaining the benefits of partially treated water, such as the dilution effect for the sewage impact, based on the flow and volume of the discharged water. The bird life has increased significantly (Madden, personal communication and site visit, 2018b; Pretorius, personal communication and site visit, 2018), "the bird life is at its best now" (De Jager, personal communication 2018a). The water can be used for agriculture; there has been a successful turnaround of crops. Therefore, the impact on food production or livestock production is minimal (Govender, personal interview, 2018), and there has not been any impact on the cattle (Blesbokspruit Forum 2018b; van Niekerk, personal interview, 2018).

Even though no detrimental impacts have occurred, and there is acknowledgement that the water quality is better than before the STT, key individuals are aware that it is an interim situation (De Jager, personal interview, 2018b; GDARD representatives, personal interview, 2019; Govender, personal interview, 2018; Liefferink, personal interview, 2016; Madden personal interview, 2018a; Maurizi, personal interview, 2018; Tlale, personal interview, 2018a; Van der Merwe, personal interview, 2018; van Niekerk, personal interview, 2018). However, concerns about the quality of the neutralised water that is discharged into the Blesbokspruit were still raised. The lack of information regarding the sulphates was a major concern, and information regarding recent trends and analysis would enable users of the water to plan for the future (van Niekerk, personal interview, 2018). Concerns stemmed from the probability that there will be nothing left in the water if the high sulphate levels continue (Van der Merwe, personal interview, 2018), "the plant is not a problem as long as they sort the salt problem out" (Madden, personal communication and site visit, 2018b). Madden, through his knowledge of the area, is aware of similar problems that occurred in the 1990s and knows what the possible impacts on the water can be. Even though the DWS is managing the problem, but whether it is acceptable or reasonable is the unanswered question (van Niekerk, personal interview, 2018), because "it cannot go on forever" (Van der Merwe, personal interview, 2018).

These concerns stemmed from what the quality of the water will be for stakeholders' own use and what impact it will have on groundwater (Digby Wells Environmental 2014; Naidoo, personal interview, 2018). Owing to the fact that the possible impacts have not been assessed, a sample of the neutralised water and a full detailed analysis, hydrological flows of qualities and ecological sensitivities would be needed to assess and determine how the discharge can impact on the quality of the water and further downstream in the Vaal Catchment (Pillay, personal interview, 2018). "The low quality of water in the Vaal barrage necessitates a periodic release of water from the Vaal Dam to reduce the salinity for the downstream Vaal River users" (Environomics 2014, p. 29). During rainy seasons this impacts on the water quality. However, drought periods might require releases from the upper Vaal River for dilution purposes, which is reserved for other uses. Rain naturally dilutes the highly saline water, whereas limited rain will require other dilution means to ensure the Vaal Barrage does not reduce its water quality further (DWS 2013, p. 29). Since the STT began in 2016, no dilution releases were required from the Vaal Dam, and it was unexpected during low rainfall and runoff periods due to the volumes discharged of up to 110 Mℓ/day (DWS 2019).

During the fieldwork conducted for this book, a consistent concern raised by key individuals was with the large volumes of water in the Blesbokspruit (De Jager, personal interview, 2018b; Madden, personal interview, 2018a). The volumes of water in the Blesbokspruit have increased due to the discharge (Maurizi, personal interview, 2018). Key individuals who expressed concerns about water levels were those living on the embankment of the Blesbokspruit or those in the tourism sector who have lost their shoreline due to excessive quantities of water and more so when it rains (De Jager, personal communication, 2018a; De Jager, personal interview, 2018b; Madden, personal interview, 2018a; Maurizi, personal interview, 2018; Pretorius, personal communication, 2018). Other key individuals who expressed concerns about the increased water levels were those who worked in positions related to the management of the Blesbokspruit, such as GDARD, and the CoE. However, Pillay suggested that without the flow of the water, the Vaal catchment would reduce (Pillay, personal interview, 2018), implying that the discharge of water (and not necessarily the neutralised water) is necessary for the system. Key individuals from the agricultural sector are more concerned about the high sulphate levels rather than volumes of water (Agricultural sector representative, personal interview, 2018; van Niekerk, personal interview, 2018). Therefore, the increased flow was an advantage to the agricultural sector but varied for the community and tourism sector when coupled with heavy rainfall, which led to flooding.

In mid-September 2018, operations at the AMD treatment plant ceased temporarily (Blesbokspruit Forum 2018d; Tlale, personal communication, 2019). This was due to a major fault in one of the pumps/control panels, which had to be replaced (Madden, personal communication, 2018c; Tlale, personal communication, 2019). The pumping resumed in January 2019 (Blesbokspruit Forum 2019b). In May 2019 it was reported that the water quality results for the plant were compared against three limits, which are the Wastewater Limits, Directive Limits and the Blesbokspruit In-stream Water Quality Guidelines (Blesbokspruit Forum 2019b). The plant was well within the Directive Limits but was slightly higher in terms of the Wastewater Limits. When it rains, the wetland is filled with cleaner water. During this time, the water level was very low because the AMD treatment plant was not pumping and discharging the water. This meant that the environment is reforming itself. However, when the plant starts pumping, the salinity levels in the water will again be high (Madden, personal communication, 2018c).

While the plant stopped pumping, a situation arose involving the Cowles Dam[10] turning pink in December 2018, which was communicated through the Blesbokspruit Forum Google group (Blesbokspruit Forum Google group 2018a, b). Some stakeholders suggested that it was a plume that appeared in August 2018 and was of the opinion that it resulted from poor water quality containing unidentified bacteria, and possibly a salinity issue (Blesbokspruit Forum 2019a). Liefferink (Blesbokspruit Forum Google group 2018a, b) suggested that the colour indicates that it was caused by a sewage spill. This could mean that, just like rain, the discharged water from the treatment plant provides a necessary dilution process of the polluted water. This problem first occurred as far back as 2010, when Grootvlei Mine was charged with discharging untreated water into the Blesbokspruit, shortly after, pumping stopped and it became easier to identify other forms of pollution in the river (Blesbokspruit Forum Google group 2018a, b). Thus, the discharge of water from the AMD plant could be assisting to lessen other impacts on the Blesbokspruit. In February 2019 a representative of Rand Water reported that the AMD treatment plant had a high impact on the flow of the Blesbokspruit and high sulphate levels, with the back-end discharge of AMD resulting in higher sulphates levels. However, the situation normalised once the plant was operational again. Therefore, "the plant is proving to be good for the system" (Blesbokspruit Forum 2019a). However, concerns around the EIA were unresolved, creating negative perceptions of the AMD STT based on the EIA process.

5.6.3 Perceptions of the Significance of an EIA Process

> *I find the plant interesting, exciting and frightening, because there are aspects that haven't been looked at.* (Humphreys, personal interview, 2018)

When stakeholders became aware of the sludge being disposed of in the mine void, they raised many questions and concerns about the actual STT, because the purpose of an EIA was to identify possible risks to the people and the environment. Without an EIA for the mine void and the discharged water, stakeholders were concerned about the quality of the water. People have an emotional attachment to the water due to their personal interests, and unknown impacts of the STT could jeopardise their use of the water, creating disagreements over how the resource is managed and controlled. Stakeholders felt that the DWS had exerted its power over the public through the implementation of the STT. The public used their local knowledge in order to affirm their own identity, and resist the effects of this power and questioned the efficiency of the process. Various stakeholders have an attachment with the water of the Blesbokspruit, and this leads to conflict with the forms of water governance and the power relations embedded in the management of this resource. In this way, water users developed trust issues with the DWS and the TCTA.

[10] The Cowles Dam is in Springs.

A few meetings with different communities in different areas took place during the EIA process over a period of almost 2 years. These meetings were quite intense (Pillay, personal interview, 2018), because stakeholders became anxious about the STT and its impact. Maurizi (personal interview, 2018), a community member and employee at the CoE, said that it was not often that a public interest project was completely changed to consider the public's views. Many people who attended the public meetings for the EIA process were ordinary residents with no specific knowledge. However, there were also people who were extremely knowledgeable about the STT process who thought of things that the project team had not, and their options should have been explored, even if the decision involved changing the original plan because the suggestions were valid (Maurizi, personal interview, 2018).

Maurizi (personal interview, 2018) referred to the suggestions that the public (IAPs) made during the EIA public participation process to use the mine void. Residents asked the right questions: "We asked the questions about whether the sludge can be put underground and they said no there are issues with that, and the next minute they are putting it underground" (Maurizi, personal interview, 2018). However, according to Tlale (personal interview, 2018a) and Pillay (personal interview, 2018), at that stage no one in the project team thought it would be feasible to use the mine void; the option was never considered initially or assessed. This led Maurizi to the view that projects were implemented depending on who the project team was and not what the best option was. She said due to this the process should take into account what the public suggested and asked because that, essentially, was the purpose of an EIA public participation process. However, this did not happen. People asked questions based on their knowledge of the area and their emotional attachment to the water, even though they may not have been groundwater experts. Their questions should have been investigated (Maurizi, personal interview, 2018).

Further, the significance of the EIA process is to identify the long-term implications, but now that an EIA was not conducted for the mine void, no one knows what will happen and whether the public comments were heard and action was taken (Barker, personal interview, 2018). Barker and Maurizi both expressed how the public's voices and suggestions were not heard, which defeated the meaning of public participation. If the public's suggestions had been considered and investigated at the EIA stage, risk assessments could have been conducted for the use of the mine void. However, according to Tlale (personal interview, 2018a), the TCTA would not have continued conducting the EIA for the Grootvlei Site 6/L/16 if they thought of using the mine void. However, Maurizi (personal interview, 2018) questioned what the purpose of an EIA was, if not to consider the options presented. She explained the different types of scenarios: a bad one, a very bad one and the worst one and further took into account the time constraints that depend on what the scope of work allows for. The problem is that if new options arise, they will delay the process and she does not know whom it is worse for: the consultant (the TCTA and Digby Wells Environmental) or the owner of the project (the DWS). Perhaps while the EIA was taking place, the project team did not want to add anything new to the project and examine all the options or to consider that there was a more suitable site. This is the reason for the phases in the EIA process. Many valuable suggestions were made during the EIA public participation process, considering the fact that many of the stakeholders present were experts in their field. The use of the mine void had already been suggested in 2014/15, but was not considered. For this reason, the research found that AMD treatment had gained a 'bad reputation'. This is not to assign blame, but to indicate reasons why stakeholders have negative perceptions of the STT.

Stakeholders took the EIA process seriously and made valuable contributions but felt that they were wrongly informed and unintentionally misled by the authorities. "The fact that we have to deal with different environmental practitioners and half of them are dishonest. They have to tell us whatever their client [the DWS] tells them, because they're getting paid for it. We can go there with thousands of complaints and queries but people don't get proper answers. Our concerns aren't really taken into account" (De Jager, personal interview, 2018b). The dynamic involved is dealing with environmental consultants who do not have objective views but are given the task of leading the process (Barker, personal interview, 2018). A perception of a solution is being created but, in fact, the LTT is not recognised by the majority of them; as long as the immediate satisfaction and need are addressed and the "rules" are followed, one cannot object (Barker, personal interview, 2018). Some of the objections and contributions came from people who have lived in the area for many years and understand the complicated dynamics, "these people were very knowledgeable" (Maurizi, personal interview, 2018). Keeping people out of the discussion or misleading them results in negative consequences. When people's views are rejected, it indicates that they are seen as unknowledgeable (Zwartaveen 2015). This seemed to be the case during the EIA public participation process as some concerns raised were not answered by the environmental consultant. Further, when questions were asked in 2018 in the Klip River Forum, which is a public participation forum, similar questions still could not be answered due to the topic being 'confidential' (Klip River Forum 2018).

The issue of borehole/groundwater was a common concern among community members. They use this water – which they view as being of good quality – for drinking, cooking and other domestic purposes; "we are solely dependent on groundwater which sustains both human and animal life" and AMD will at some stage ruin the groundwater (De Jager, personal interview, 2018b; Digby Wells Environmental 2014). Some community members provided examples of how boreholes were lost due to the negative impacts of AMD, which indicated that they already had a perception and fear that the STT would further exacerbate the problem. They were concerned that their source of water would be affected and contaminated (Digby Wells Environmental 2014). However, they were informed that a monitoring plan would be in place and that the sludge disposal facility contains a leak detection system in the event that the sludge infiltrates the groundwater. When the STT launched, the public slowly became aware that the mine void was being used for the sludge disposal. This explains why they felt negative about the STT, which was based on the EIA process, and also because their views had not been taken into consideration. It also explained negative sentiment about the mine void, because when it was suggested during the public participation process, it was overlooked and considered to be flawed due to possible impacts on groundwater.

Therefore, instead, to sway the public opinion towards using the preferred site (the above-ground Grootvlei Site 6/L/16 TSF) which the DWS had selected meant answers were given without realising what future implications would be brought about. The result is that the mine void that was initially considered flawed, now being used, is labelled a pilot project that resulted from emergency measures, and is exempted from an EIA.

It was found that perceptions of the STT that stemmed from the EIA process influenced people's constructions of the water quality. The public are now of the view that the mine void could impact on groundwater, as they were told during the EIA process. They are of the opinion that they have been misled in order to serve the interests of the government in a problem for which they hold the government responsible. Human-created events that occur in the environment lead to socially produced ways of experiencing and understanding the environment, indicating how power relations play a role in how water quality is constructed.

5.7 Conclusion

The perceptions of the AMD STT stem from the fact that there are many 'unknowns' regarding the STT of AMD because no EIA was conducted. Some stakeholders felt misinformed and misled due to how long it took to implement the treatment and the implementation process that was followed. Further, they felt that the DWS was responsible for the cost of treatment because the mining industry had not been held accountable for polluting the water due to poor management. Despite this, the project team tried to promote the benefits of the STT by comparing it to AMD in its original state. However, there was already lack of trust between the public and the DWS, because of the EIA process, and the feeling of being misled when the STT was implemented. The fact that the mine void is used for the sludge disposal enabled the public to raise concerns about its impact on groundwater and what the purpose of an EIA was. This meant that the public participation process had no merit because their views and suggestions were not taken into account. The transfer of information and lack of communication to the public regarding the STT led them to question the management and legitimacy of the STT, and that the decisions made were linked to power play. Even though stakeholders show awareness that the STT is more beneficial than untreated water, their perceptions stem from the unknown impacts, and the frustration of being misinformed and misled. Further, owing to their interests in water, for various stakeholders, the STT has just as much perceived implications and concerns as AMD in its original state because no environmental and social impact assessments were conducted. The purpose of this chapter was to present these perceptions. Chapter 6 discusses a range of factors that influence how individuals socially construct the water quality of the Blesbokspruit.

References

Agricultural sector representative. 2018. Personal interview, 12 April. Springs, South Africa.

Ambani, AE. 2013. *Long-term assessment of the surface water quality in the Blesbokspruit Ramsar wetland.* Unpublished MSc dissertation. Johannesburg: University of Johannesburg. Available at: https://ujcontent.uj.ac.za/vital/access/manager/Repository/uj:11619. Accessed on: 21 June 2019.

Anticevich, A. 2014. No, no to this sludge dam: Residents are fighting for the preservation of Blesbokspruit, their properties. *Springs Advertiser,* 28 July. Available at: https://springsadvertiser.co.za/96610/sludge-dam/. Accessed on: 10 January 2018.

Barker, A. 2018. Consultant: DRD Gold. Personal interview, 4 June. Panorama, South Africa.

Bega, S. 2021. Toxic mine water still a threat. *Mail & Guardian,* 28 May to 3 June.

Blesbokspruit Forum. 2017. Minutes Blesbokspruit Forum, 2 February. Available at: http://www.reservoir.co.za/forums/vaalbarrage/blesbok_forum/blesbok_home.htm. Accessed on: 31 January 2020.

Blesbokspruit Forum. 2018a. Minutes of Blesbokspruit Forum meeting and personal attendance, 8 February 2018.

Blesbokspruit Forum. 2018b. Minutes of Blesbokspruit Forum meeting and personal attendance, 11 May 2018.

Blesbokspruit Forum. 2018c. Minutes of Blesbokspruit forum meeting and personal attendance, 8 August 2018.

Blesbokspruit Forum. 2018d. Minutes of Blesbokspruit Forum meeting and personal attendance, 8 November 2018.

Blesbokspruit Forum. 2019a. Minutes of Blesbokspruit Forum meeting and personal attendance, 7 February 2019.

Blesbokspruit Forum. 2019b. Minutes of Blesbokspruit Forum meeting, 9 May 2019.

Blesbokspruit Forum. 2019c. Minutes of Blesbokspruit Forum meeting, 8 August 2019.

Blesbokspruit Forum. 2019d. Minutes of Blesbokspruit Forum meeting and personal attendance, 7 November 2019.

Blesbokspruit Forum Google group. 2018a. Cowles Dam. E-mail communication, 14 September 2018.

Blesbokspruit Forum Google group. 2018b. Cowles Dam. E-mail communication, 1 November – 6 December 2018.

Bosman, C. 2018a. Consultant for Environmental health services. Personal interview, 12 April. Pretoria, South Africa.

Bosman, C. 2018b. IWRM, the NWA and Water Use Authorisations: Focussing on WULAs, and IWWMPs. 13 – 15 March 2018. Carin Bosman Sustainable Solutions (CBSS) Training Course: Pretoria.

Budds, J. & Hinojosa, L. 2012. Restructuring and rescaling water governance in mining contexts: The co-production of waterscapes in Peru. *Water Alternatives,* 5(1):119–137.

Creamer, M. 2015. Water: Decide now on Lesotho 2, acid mine drainage projects – Prof. *Mining Weekly.* Available at: http://www.miningweekly.com/article/waterdecide- now-on- lesotho-2. Accessed on: 20 January 2016.

De Jager, A. 2018a. Community member. Personal communication, 27 February. Springs, South Africa.

De Jager, P. 2018b. Lawyer and Springs resident. Personal interview, 25 January. Springs, South Africa.

Department of Water Affairs and Forestry (DWAF). 1996. *South African Water Quality Guidelines* (2nd Edition). Volume 1: Domestic Use. Pretoria: DWAF. Available at: https://www.dws.gov.za/Groundwater/documents/Pol_saWQguideFRESHDomesticusevol1.pdf. Accessed on: 10 May 2021.

Department of Water and Sanitation (DWS). 2013. *Feasibility study for a long-term solution to address the acid mine drainage associated with the East, Central and West Rand underground mining basins: Assessment of the water quantity and quality of the Witwatersrand mine voids* (Technical prefeasibility report Study report No. 5. May 2013). Pretoria: DWS.

Department of Water and Sanitation (DWS). 2017. *Mine Water Management: Policy Position Draft for External Consultation and Discussion*. Gazette No. 658. Available at: https://www.greengazette.co.za/notices/national-water-act-36-1998-mine-water-management-policy-position-draft-for-external-consultation-and-discussion_20170707-GGN-40966-00658.pdf. Accessed on: 18 October 2018.

Department of Water and Sanitation (DWS). 2019. *National Water and Sanitation Master Plan. Volume 1: Call to Action, Ready for the Future and Ahead of the Curve*. Pretoria: DWS.

Digby Wells Environmental. 2014. *Environmental impact assessment for the construction of the proposed sludge disposal facility and pipeline associated with the treatment of acid mine drainage from the eastern basin of the Witwatersrand, Gauteng: Full comment and response report*. Johannesburg: Digby Wells and Associates.

Digby Wells Environmental. 2015. *Construction and operation of the proposed sludge disposal facility and pipelines associated with the treatment of acid mine drainage in the eastern basin of the Witwatersrand, Gauteng: Final environmental impact assessment report*. Johannesburg: Digby Wells and Associates.

Dirisu, C. & Mafiana, M. 2016. Level of pH in drinking water of an oil and gas producing community and perceived biological and health implications. *European Journal of Basic and Applied Sciences*, (3):3. Available at: https://www.researchgate.net/publication/332012834_level_of_ph_in_drinking_water_of_an_oil_and_gas_producing_community_and_perceived_biological_and_health_implications. Accessed on: 14 august 2020.

Environomics. 2014. *Gauteng environmental management framework report*. Johannesburg: Gauteng Province, Agriculture and Rural Development. Available at: http://www.gauteng.gov.za/government/departments/agriculture-and-rural-development/Documents/Gauteng%20Provincial%20Environmental%20Management%20Framework/2014%20GPEMF%20Section%20A%20Introduction%20and%20Status%20Quo.pdf. Accessed on: 20 June 2019.

Fourie, B. 2019. Aquatic Ecologist and Consultant: City of Ekurhuleni (CoE). Personal interview, 8 February. Midrand, South Africa.

GDARD Representatives (Jonhasi, C. Maluleke, R., Joshua, Q. & Mashau, B.) 2019. Gauteng Department of Agriculture and Rural Development (GDARD). Personal interview, 5 March. Marievale Bird Sanctuary, South Africa.

Govender, B. 2018. Chief directorate Mine water management unit: Department of Water and Sanitation. Personal interview, 1 March. Pretoria, South Africa.

Holden, R. 2019. Business Analyst: TCTA. Personal Communication, 12 March.

Humphreys, J. 2018. Ekurhuleni Ward Councillor: Bedfordview. Personal interview, 11 April. Nigel, South Africa.

Klip River Forum. 2018. Minutes of Klip River Forum meeting and personal attendance, 8 May 2018.

Kolver, C. 2014. Govt lauded for AMD paradigm shift, but complexity not fully grasped. *Mining Weekly,* 18 April. Available at: http://www.miningweekly.com/article/govt-lauded-for-amd-paradigm-shift-but-detractors-insist-it-hasnt-fully-grasped-complexity-of-the-issue-2014. Accessed on: 7 October 2019.

Liefferink, M. 2016. CEO: Federation for a Sustainable Environment. Personal interview, 21 November. Bryanston, South Africa.

Madden, S. 2018a. Environmentalist and Springs resident. Personal interview, 8 February. Springs, South Africa.

Madden, S. 2018b. Environmentalist and Springs resident. Personal communication and Ramsar site visit, 27 February. Springs, South Africa

Madden, S. 2018c. Environmentalist and Springs resident. Personal Communication, 13 December 2018.

Maurizi, A. 2018. Manager: Water Quality Division, Ekurhuleni. Personal interview, 3 May. Springs, South Africa.

Meyer, E. 2018. Blesbokspruit Trustee Member. Personal interview, 2 February. Springs, South Africa.

Mpofu, C., Morodi, TJ. & Hattingh, JP. 2018. Governance and socio-political issues in management of acid mine drainage in South Africa. *Water Policy*, 20:77–89. Available at: https://iwaponline.com/wp/article/20/1/77/38136/Governance-and-socio-political-issues-in. Accessed on: 1 June 2020.

Naidoo, S. 2017. *Acid mine drainage in South Africa: Development actors, policy impacts and broader implications*. Switzerland: Springer Nature.

Naidoo, P. 2018. Manager Platinum and Metals: Impala Platinum. Personal interview, 7 March. Springs, South Africa.

Pillay, M. 2018. Business Development Executive: Digby Wells Environmental. Personal interview, 16 January. Bryanston, South Africa.

Pretorius, B. 2018. Owner at Bush Inn and Stable Inn conference venue. Personal communication and site visit, 10 May. Springs, South Africa.

Reservoir website. 2021. Water quality guidelines. Available at: https://www.reservoir.co.za/forums/vaalbarrage/blesbok_forum/blesbok_documents/BF_WQGuidelines.pdf. Accessed on: 10 May 2021

Sicetsha, A. 2020. Khulubuse Zuma, Zondwa Mandela scot-free after NPA drops charges. *The South African,* 26 February. Available at: https://www.thesouthafrican.com/news/khulubuse-zuma-zondwa-mandela-scot-free-after-npa-drops-charges/. Accessed on: 10 October 2020.

Springs Advertiser. 2014a. Blesbokspruit gets too much water: The nutrient-rich water caused the reeds to overgrown the spruit. *Springs Advertiser*, 13 April. Available at: http://springsadvertiser.co.za/89328/blesbokspruit-gets-too-much-water/. Accessed on: 22 February 2017.

Springs Advertiser. 2014b. Springs says no to opencast mining: Residents oppose these-zoning of the land. *Springs Advertiser*, 23 April. Available at: https://springsadvertiser.co.za/90116/springs-says-opencast-mining/. Accessed on: 22 February 2017.

Springs Advertiser. 2014c. "Water from plant may not be safe: Water from the acid mine drainage treatment plant, now being built, will be discharged into Blesbokspruit" *Springs Advertiser* 21 October 2014. Available at: https://springsadvertiser.co.za/103416/water-from-plant-may-not-be-safe/. Accessed on 10 January 2018.

Springs Advertiser. 2014d. "What about our Blesbokspruit?" Springs Advertiser 4 November 2014. Available at: https://springsadvertiser.co.za/104050/what-about-our-blesbokspruit/. Accessed on 10 January 2018.

Springs Advertiser. 2015. Community's concerns about sludge disposal facility addressed. *Springs Advertiser*, 30 June. Available at: https://springsadvertiser.co.za/122293/communitys-concerns-about-sludge-disposal-facility-addressed/. Accessed on: 10 January 2018.

Tlale, S. 2018a. AMD project manager: TCTA. Personal interview, 9 March. Centurion, South Africa.

Tlale, S. 2018b. AMD project manager: TCTA. Personal communication and AMD Treatment Plant Site Visit, 09 April 2018. Springs, South Africa.

Tlale, S. 2019. AMD project manager: TCTA. Personal communication, 11 April.

Turton, A. 2018. Mine voids. Personal communication, 18 October 2018.

Van der Merwe, C. 2018. Chairman: Blesbokspruit Trust. Personal interview, 21 February. Springs, South Africa.

Van Niekerk, J. 2018. Consultant: Karan Beef. Personal interview, 15 February. Roodepoort, South Africa.

Van Zyl, B. 2016. General Manager: Transvaal Agricultural Union (TLU). Personal interview, 11 October. Pretoria, South Africa.

Zwarteveen, M. 2015. *Regulating water, ordering society: Practices and politics of water governance*. Amsterdam: UNESCO Institute for Water Education, Universiteit van Amstar.

Chapter 6
Factors Influencing Social Constructions of the Water Quality of the Blesbokspruit

Abstract The Blesbokspruit is a large wetland system with many activities (e.g. agriculture, mining and wastewater treatment works) taking place in the catchment that contribute to the increased volume of discharged water into the river. Key individuals are aware of how the physicality of the river has changed as a result of mining and other activities that occurred over the years. There is consensus amongst these individuals that even though the Blesbokspruit was co-produced by the mining industry, it is also responsible for the deterioration of the water quality. Now other discharges, such as those from the acid mine drainage (AMD) short-term treatment (STT), have contributed to the increased volume of water. However, unknown impacts of the STT make further deterioration of the water quality possible. The way individuals attribute meaning to water tends to depend on how they value its use and what their interest in the water are. In the case of the Blesbokspruit, these factors include: (1) key individuals' understanding of the historical context of mining, (2) coal-mining activities and its impact on agriculture, (3) flows and volumes of water in the Blesbokspruit catchment, such as the impact of sewage, reed encroachment and flooding, (4) the use of technology and various processes and its impact on water quality, (5) information and communication about the AMD STT and (6) the vested interests of key individuals based on their need and use for water. These factors are discussed in this chapter to indicate how they are not self-evident or 'natural' but influence the key individuals in their social constructions.

Keywords Coal mining · Social constructions · Water quality · Neutralised water · Vested interests

6.1 Introduction

> *I saw this Blesbokspruit and I immediately became fascinated with it and it was a vastly different site to what it is today.* (Madden, personal interview, 2018a)

S. Naidoo, *Social Constructions of Water Quality in South Africa*,
https://doi.org/10.1007/978-3-030-98237-9_6

> *You don't notice the damage in the Blesbokspruit from afar, but it's struggling: It's an old lady that's been rehydrated for the meantime.* (Storey, personal interview and site visit, 2018)

The Blesbokspruit is a large wetland system with many activities (e.g. agriculture, mining and wastewater treatment works) taking place in the catchment that contribute to the increased volume of discharged water into the river. Key individuals are aware of how the physicality of the river has changed as a result of mining and other activities that occurred over the years. There is consensus among these individuals that even though the Blesbokspruit was co-produced by the mining industry, it is also responsible for the deterioration of the water quality. Now other discharges, such as those from the acid mine drainage (AMD) short-term treatment (STT), have contributed to the increased volume of water. However, unknown impacts of the STT make further deterioration of the water quality possible.

Key individuals[1] are familiar with the Blesbokspruit and are therefore aware of the known sources of pollution that have occurred over the years. This familiarity which stems from having lived or worked in the area for decades and a range of other factors identified by the research has an influence on how these individuals construct the water quality. Mining attracted people to the area for employment for instance, but because mining has caused extensive environmental degradation that required rehabilitation (e.g. AMD STT), it also influenced how key individuals construct the water quality. This suggests that vested interests, among other factors, have a strong influence. The purpose of this chapter is to identify and explain the main factors that have influenced individuals' constructions of the water quality.

The way in which individuals attribute meaning to water depends on how they value its use and what their interest in the water is. In the case of the Blesbokspruit, these factors include (1) key individuals' understanding of the historical context of mining; (2) coal mining activities and its impact on agriculture; (3) flows and volumes of water in the Blesbokspruit catchment, such as the impact of sewage, reed encroachment and flooding; (4) the use of technology and various processes and its impact on water quality; (5) information and communication about the AMD STT; and (6) the vested interests of key individuals based on their need and use for water. These factors will be discussed to indicate how they are not self-evident or "natural" but influence the key individuals in their social constructions.

6.2 Historical Context of Mining

> *Without [the mining] industry, the Blesbokspruit would not look the way it does now.* (Meyer, personal interview, 2018)

[1] In this book, the concept stakeholders refer to individual persons, organisations, government institutions, civil society, the business sector and other organisations with an interest in the Blesbokspruit. Key individuals were interviewed for this research and are persons among these stakeholders who play a prominent role in, or have specialised knowledge on, the Blesbokspruit area regarding AMD treatment. Therefore, this book refers to both stakeholders and key individuals.

The specific focus of this section is on the physical changes that occurred in the Blesbokspruit due to mining's historical impact on the water and water quality. The history of mining contributed to key individuals' understanding of how water is social, meaning that their understanding of the water quality corresponds to something in the real world, indicating that social constructions can be the product of historical events and the way such events are interpreted. AMD is caused by the mining industry's neglect in treating water adequately before discharging it into the Blesbokspruit. This is based on key individuals' knowledge of the impact of mining on water quality and agricultural land. The purpose of this section is to explain why or how the perceptions of AMD STT are based on how their knowledge or experience of mining in a historical context influences their constructions of the water quality of the Blesbokspruit. This will be illustrated by referring to the perceptions of the key individuals.

When gold mining boomed on the East Rand (discussed in Sect. 3.3), it contributed extensively to the country's economic growth (Van der Merwe, personal interview, 2018). When mines were in their prime – Marievale Consolidated Mine and Grootvlei Proprietary Mines were newly established at the time – it was a massive development period. At the time Anglo American owned and managed much of the land that was mined by different mining companies, such as Daggafontein Mine (De Jager, personal interview, 2018b; Madden, personal interview, 2018a; Van der Merwe, personal interview, 2018). However, large portions of this land were undermined (Madden, personal communication and site visit, 2018b), meaning there were large areas of land available for other uses. During the 1960s, as mining companies closed down and Anglo American no longer needed the land, they donated 750 ha, known as the *Anglo Reserve*, located within the Ramsar site area, to the Springs City Council (now called the City of Ekurhuleni (*CoE*)) with a sum of money (in the millions of rands) as a start-off to use for other developments (De Jager, personal interview, 2018b; FSE 2017; Van der Merwe, personal interview, 2018). The aim was to incorporate this land into the Marievale Bird Sanctuary. The Gauteng Department of Agriculture and Rural Development (GDARD) are the custodians and responsible for controlling the land that covers the sanctuary. The rest of the area, including the Grootvaly Educational Centre situated along the Welgedacht road, is privately owned (Madden, personal interview, 2018a).

As mines began to close down in the 1960s, new environmental legislation regarding rehabilitation was already in place. However, this was not the case prior to this period when Anglo American was done with its business operations on the East Rand (Madden, personal communication and site visit, 2018b); "Anglo American is a bit of a disgrace" (De Jager, personal interview, 2018b). Van der Merwe[2] (personal interview, 2018) concurred by indicating that Anglo American had made millions of rands from mining on the East Rand. However, once its business was completed, it pulled out

and gave the land away to the city council to avoid incurring the environmental problems and rehabilitation costs, and the council accepted this land instead of insisting that the mining house rehabilitate it. This meant that a large environmental

[2] C. van der Merwe is a Springs community member, representative on the Blesbokspruit Forum and the Chairperson of the Blesbokspruit Trust.

burden fell on the state, with the council not realising the implications that came with the poor condition of the land before accepting it. Further, illegal mining, such as gold panning, started taking place (Van der Merwe, personal interview, 2018), and other changes in the environment began to occur (Madden, personal interview, 2018a). This included the pumping of untreated water in 1995 from the Grootvlei Mine into the Blesbokspruit. Not all mining companies practised ethical mining, unlike Daggafontein Mine, which can be referred to as a great legacy of mining (Madden, personal interview, 2018a). This is based on the good condition in which the mine was left.

The modification of the Blesbokspruit system was ongoing, indicating that people cannot escape the physical changes because for that period it is fixed, and "we cannot pick it up and put it somewhere else" (Madden, personal interview, 2018a). Madden[3] (personal interview, 2018a) attributed these changes to the East Rand being one of the most densely populated areas; housing and industries have put tremendous strain on the river as a drainage system. The Blesbokspruit was a stream that at one point one could literally jump over (Govender, personal interview, 2018). As a result of various impacts from various activities, the Blesbokspruit can no longer be described as a wetland but a pouring river (Humphreys, personal interview, 2018). It is basically an artificial river (Govender, personal interview, 2018). The Blesbokspruit no longer runs dry during the winter but it contains large volumes of water all year round, becoming a huge drain for all activities requiring water discharges, especially mining (Madden, personal interview, 2018a). As these developments took place, the topography and ecology of the area also started changing (Pillay, personal interview, 2018). This altered the character and biodiversity in the river ecosystem enormously, because very specific biodiversity is found in a wetland that cannot live in a river (Humphreys, personal interview, 2018). The difficulty to adapt to the changes that mining activities brought was evident when the Blesbokspruit was placed on the Montreux Record due to the decline in water quality caused by the Grootvlei Mine. Since then, the Blesbokspruit has remained on this record and continues to be a threatened site (Madden, personal interview, 2018a; Meyer, personal interview, 2018). This means that the changes are human-induced. The impact of people is too extensive for nature to restore the earth to a balance. Nature can balance itself but the intensity of human activity is widespread, which resulted in a deteriorated system (Van Zyl, personal interview, 2016).

There is a common awareness among key individuals that it is the pumping of underground mine water that created the Blesbokspruit (De Jager, personal interview, 2018b; Govender, personal interview, 2018; Madden, personal interview, 2018a; Pillay, personal interview, 2018; Van der Merwe, personal interview, 2018). The increased water volume in the Blesbokspruit is attributed to the ongoing pumping rates during the predominant years of mining: in the 1950s, from 93 Mℓ/day to the 1980s–1990s at 63 Mℓ/day (Scott 1995, p. 33) to the current AMD treatment

[3] S. Madden worked in the mining industry for over 40 years. He is a well-known environmental activist in Springs. He placed Marievale Bird Sanctuary on the map by helping to ensure that larger portions of the land remain protected for the bird sanctuary; therefore he is also linked to the tourist sector.

plant pumping between 80 Mℓ/day and 110 Mℓ/day since 2016 (Tlale, personal interview, 2018). Pumping caused the wetland to widen and created a significant influence on the system (Madden, personal communication and site visit, 2018b; Pillay, personal interview, 2018). Therefore, "without [the mining] industry, the Blesbokspruit would not look the way it does now" (Meyer, personal interview, 2018). There was permanent water in places and a benefit of this is that it attracted birds to the area. Birds were not present in large numbers prior to this (Pillay, personal interview, 2018). This artificial nature of the river is further attributed to wastewater treatment works and the discharge from the AMD treatment plant contributing to increased volumes of water (Blesbokspruit Forum 2018d; Govender, personal interview, 2018; Labuschagne, personal interview, 2018).

The fact that the volume of water in the river proved to be significant to the system also had negative consequences. In the late 1990s mining became so extensive that it contributed immensely to the poor water quality, and then again when mining ceased in 2011, it became problematic for the community. This was because the water was no longer pumped and the risk of AMD seeping into surface water was concerning. However, even though concerns of AMD were raised, the water quality was said to have improved after mining activity had ceased (Madden, personal interview, 2018a). Even though the mining industry has contributed significantly to the deteriorated water quality, AMD did not breach as it did in the western basin (Meyer, personal interview, 2018). According to Meyer,[4] reasons for the improvements "from an inorganic chemical pollution point-of-view" include existing mines and associated industries employing environmental scientists and consultants whose duties are defined by the laws of the country. This resulted in a more responsible attitude towards the environment in a number of mining operations. The result is that mines discharge "cleaner" water into the Blesbokspruit as compared to in the past. In addition, several industries along the river have closed down due to economic circumstances (Meyer, personal interview, 2018), and since most mining activity ceased, the water quality has improved.

The East Rand became attractive to the people who moved there because of employment opportunities, and they were fully aware of the mining activities that took place (Pillay, personal interview, 2018). Pillay[5] is of the opinion that how the community views the environmental problems boils down to what is feasible for them. The views of gold mining changed based on the increasing socio-economic impacts on the public. Community members felt that they were pushed around by the mining industry, as the government favoured the mining industry over the environment, "the community has no confidence in the government" (Van der Merwe, personal interview, 2018). "[Even though] environmentalists were beating their drums for the last 10 to 15 years nothing was done" (Naidoo, personal interview, 2018). Therefore, people have suffered tremendously due to the environmental impacts caused by mining over the years (Madden, personal interview, 2018a).

[4] E. Meyer was a resident of Springs for many years. However, despite having moved out of the area, he still remains a member of the Blesbokspruit Trust.

[5] M. Pillay is an environmental consultant at Digby Wells Environmental, an environmental consulting firm, and worked on the EIA for the sludge disposal for the STT of AMD.

However, Pillay (personal interview, 2018) believed that appreciation should be given to the mining sector because that is how the communities were created. Further, the area became attractive to tourists because of the Marievale Bird Sanctuary, as species of birds diversified due to the increased water in the Blesbokspruit. As in the case of the East Rand, many cities in the world have many recreational areas being transformed into economic opportunities.

People who have lived on the East Rand for most of their lives have been at the centre of mining impacts on water; "mining made South Africa, but it could also destroy it" (De Jager, personal interview, 2018b). The Grootvlei Mine is an example of such impacts. Ownership of the mine changed several times and the last owners contributed to severe impacts on the water quality (as explained in Sect. 4.3). What was left was an abandoned site, slimes dams were not maintained, and dust had an impact on the people. According to Madden (cited in FSE 2017), the East Rand is one of the finest areas in Gauteng for farming. "The horror of mining" that companies such as Aurora Empowerment Systems (the Grootvlei Mine owners) caused led to the Springs community having no faith in mining and after all the damage "now they want coal too". The knowledge of historical mining and its impacts have led community members to oppose new mining activities.

Coal mining has been in existence for many years since the 1940s. However, from the 1960s, coal mines have begun to close and become abandoned. New opencast coal mining applications during 2018 caused debates among various stakeholders (Blesbokspruit Forum 2018a, b, c, d). Due to the East Rand having to undergo excessive rehabilitation, such as AMD STT, after years of non-compliance of environmental regulations by mining companies, the public display enormous scepticism when it comes to new mining activities. Community member, Stan Madden, who came to the East Rand to work at Marievale Consolidated Mine, and now receives his pension from this industry, had over 30 years of experience in mining (Madden, personal communication and site visit, 2018b). Madden expressed gratitude to the mining industry for his years of employment, but he could not let the environmental degradation continue. He made a decision based on what would benefit the environment and the people who depend on it. Madden (personal interview, 2018a) believed that he had a social responsibility to protect the environment. He explained:

> *I am a layman; I am not an academic, I am not a scientist, but I have a very deep feeling for nature. I am one of these very old-fashioned people who thinks that I was put here to be a custodian. We haven't done a very good job of that. I am not proud of our achievement in being custodians of nature. I am in my 91st year. I have children and I have great grandchildren and my problem, my serious worry, is that I can't show them what I have seen on this Blesbokspruit. That is to me a tragedy and it upsets me greatly that those children haven't seen what I have been able to see. That is why I am so obsessed with this, and so upset when there is a definite and blatant disregard for its health.*

Madden's construction of the water quality was based on his experience of the history of mining in the area and being exposed to both the mining industry's benefits and downfalls. His decision to take a stand was based on the livelihood he created, indicating that protecting the environment is more important than having to fix the long-term problems left behind by the mines. Such an example comes from one's understanding of the impacts of mining on water, which therefore comes from one's experience.

In retrospect, key individuals' knowledge of years of unethical mining contributed to how they viewed mining and its contribution to the physical changes in the Blesbokspruit and the impact on the quality of the water. Common to the key individuals' understanding is that the physicality of the Blesbokspruit changed due to mining's influence. The volumes of water created a change in the physicality of the river, and this change has gained both positive and negative perceptions. For instance, it positively impacted on the bird life due to the changes in the landscape of the area. However, it gained negative perceptions based on the government's lack of responsibility to hold the mining sector accountable for damages caused during and after mining, such as the pumping of untreated water. The formation of these constructions depended on what was feasible for that individual and this is based on the individual's particular interests that stem from his or her knowledge of the historical context of mining. Of importance to note is how interests have changed, for example, from interests in mining due to employment needs to environmental interests, such as the increase in bird species.

This discussion demonstrated that the history of mining is not fixed based on facts but on how the management of the water occurs, such as how the neutralised water, as a result of the STT, is discharged. The way the water is managed influences individual interests, and these interests are influenced by one's understanding of the historical context of mining and its impact on water quality. Should mining occur in future, knowledge of the history of experiences indicates the potential future impact, as in the case of new coal mining applications, which is discussed next.

6.3 Current Coal Mining

> *This is prime agricultural land and this is where they want to mine. It's a "no no". We have a legacy of mining here that is not nice.* (Madden, personal communication and site visit, 2018b)

How one understands the historical context of mining means that new coal mining activities influence how one socially constructs the water quality of the Blesbokspruit, which is discussed in this section.

With 95% of gold mines no longer in existence, there is now a big drive for coal mining, and this means that financial provision will be needed for rehabilitation and closure, which could have a direct consequence for AMD treatment (Barker, personal interview, 2018; Pillay, personal interview, 2018). The drive for future coal mining is attributed to a lack of knowledge about its impact; people treating water as infinite, but what people have failed to realise is that there will be a lot of water, but it will not be drinkable due to its excessively polluted state (Van der Merwe, personal interview, 2018). "What people don't understand is that people like us [community] that actually see under the skin, others drive past and do not know the real problems [coal mining] like we [community] do, we rely on the system" (Storey, personal interview and site visit, 2018). Coal causes significant damage to the landscape and coal mining influences subsurface water (Van der Merwe, personal interview, 2018), leading to a definite impact on the quality of water (Storey, personal interview and site visit, 2018).

During a site visit with Nigel community member Adrian Storey (see Fig. 6.1), he explained his observations of the impact of coal mining companies and is aware of the offences committed. The drainage system in the East Rand is poor, and when it rains the residue from coal mining washes straight into the Blesbokspruit, affecting the water quality. Additional coal mining initiatives have the potential to heavily increase the pollution of an already heavily polluted landscape (Meyer, personal interview, 2018). Water is already scarce and people have to decide what is more

Fig. 6.1 Adrian Storey explaining the impact of coal mining on the Blesbokspruit. (Photograph by researcher)

important, coal, a common resource (but also a dirty energy resource), or water, a scarce resource. From a community member's perspective such as that of Stan Madden, who has 30 years of knowledge from working in the mining industry, there is coal underlying the whole area that the mining industry wants to extract. There is a legacy of mining on the East Rand and the community opposes the continuation of mining, especially open-cast coal mining, which cannot take place on the door-step of so many communities due to the environmental burden caused (De Jager, personal interview, 2018b; Madden, personal communication and site visit, 2018b). "We have to balance this situation, we have reached a stage where enough is enough" (Madden, personal interview, 2018a).

Years of mining have created the challenges of AMD, but, in addition, coal mining is also known to have a detrimental impact on land that can potentially be used for agriculture. Coal mining has a lifespan of only 40 years, and once completed, a big open hole remains, meaning the land cannot be used; nothing can grow there. Therefore, agriculture will suffer, and communities have tried to prevent this because the water quality in an open-cast pit from coal mining is problematic (De Jager, personal communication, 2018a). If the entire Blesbokspruit looked like that, then there would be major problems, implying that this could happen if measures are not taken (Agricultural sector representative, personal interview, 2018).

The East Rand has extensive agricultural land split by the Blesbokspruit and the Grootvlei Mine, with the AMD STT plant visible in the distance (see Fig. 6.2). Madden indicated that coal mining companies want to use this land. Even though this land is meant for agriculture, community members such as Stan Madden are of the opinion that "if you mention mining to the government, it takes precedence over everything else" (Madden, personal communication and site visit, 2018b). He argued that the quality of the coal was so poor and cheap that it was not worth risking agricultural land or the environment. In Madden's (personal interview, 2018a) view, the East Rand has high-potential agricultural land with the highest yield (see Fig. 6.3) that can feed large numbers of people. It should therefore not be compromised by open-cast coal mining.

There is open agricultural land available on the East Rand, and due to the increase in open-cast coal mining applications, concerns were raised about how land that could be used for food production would be affected by coal mining (Madden, personal communication and site visit, 2018b; Storey, personal interview and site visit, 2018). There is a farm in Nigel that produces a harvest on about 10 000 ha of land per year, without using chemicals in the process (or genetically modified seeds or plants). Nigel community member, Adrian Storey, felt that destroying agricultural land by allowing mining to expand would be extremely detrimental to food security. Some areas are already black from coal, creating a layer of polluted air above the agricultural land (see Fig. 6.4). Even though agriculture is also known as a polluter of water, the mining industry has caused visible impacts on the environment but agriculture is seen as the "lesser of two evils". More so because agriculture provides resources that can sustain human beings. However, mining has created economic wealth and employment. For example, a residential property owner would oppose a coal mining activity due to the potential environmental impacts, but a person living in a township in the same area would back the coal mining initiative because of the potential for employment.

Fig. 6.2 Agricultural land in Springs. (Photograph by researcher)

Fig. 6.3 Agricultural land in Nigel. (Photograph by researcher)

Fig. 6.4 Coal trucks passing through agricultural land in Nigel. (Photographs by researcher)

According to Storey (personal interview and site visit, 2018), there are thousands of unemployed mining professionals because the industry is declining. Due to this decline, the responsible government department (the DMR) does not enforce legislation and is in favour of mining operations (Storey, personal interview and site visit, 2018). Such lack of implementation of environmental regulation is why mining has brought, and continues to bring, many environmental concerns, evident in the abandoned mines deteriorating in the form of AMD. Nigel resident Adrian Storey has 15 years of knowledge of the area and said he knew when a mine was not pumping and treating the water accordingly. He is of the view that there are well-protected business people who do not abide by environmental regulations. Outspoken community members, such as Adrian Storey and lawyer Philip de Jager, go to great lengths to inform government officials from the Department of Mineral Resources (DMR) and the Department of Water and Sanitation (DWS) of irregularities in mining operations. They go as far as accompanying government officials to the sites to point out evidence that the mine is non-compliant. However, action is not always taken which causes frustration for these community members (Madden, personal interview, 2018a). For this reason, Mashau (GDARD representative, personal interview, 2019) believes that increased coal mining applications were the next big war for which they are preparing themselves, "coal mining will set us back as we strive towards rehabilitating and maintaining the area". "Anything associated with coal mining is distressing"; there are possible health-related impacts, more so with coal mining than AMD (De Jager, personal interview, 2018b). According to De Jager (personal interview, 2018b), informing the Trans-Caledon Tunnel Authority (TCTA) of non-compliance from a mine is the best route to ensure action because the non-compliance of the mining operation can affect the STT of AMD and further exacerbate AMD-related problems.

New coal mining activities do not involve sustainable methods, causing further contamination to the Blesbokspruit (De Jager, personal interview, 2018b; Storey, personal interview, 2018; Van der Merwe, personal interview, 2018). Coal mining situated near the Blesbokspruit will discharge water into the spruit, impacting on surrounding users, such as Karan Beef Feedlot, which is situated further downstream. The discharged water from coal mining can contribute to increased levels of sulphates already coming from the AMD treatment plant and this will be a disaster (van Niekerk, personal interview, 2018). The high levels of sulphates are already indicated in the water quality samples conducted by Rand Water and the DWS (Meyer, personal interview, 2018). The Springs and Nigel communities are therefore opposed to new coal mining activities on the East Rand. According to them, owing to the history of unlawful activities or non-compliance with environmental regulations by coal mines, they stand together with environmental activists and ward councillors and take matters into their own hands to appeal for the rejection of new mining applications (DA Gauteng 2017; De Jager, personal interview, 2018b; Labuschagne, personal interview, 2018; Madden, personal interview, 2018a; Storey, personal interview and site visit, 2018; Van der Merwe, personal interview, 2018). This is illustrated in the case of Canyon Coal's Palmietkuilen Mine, where residents appealed against the mine conducting its operations (see Fig. 6.5). However, the decision remained out of their control. The DMR had the final decision on issuing the mining right (Naidoo, personal interview, 2018).

AMD is a well-recognised phenomenon in the East Rand and the AMD treatment plant is currently treating large volumes of AMD per day, which could have been compromised by the Palmietkuilen operation by contributing to rising salinity levels in the Blesbokspruit. Residents were not happy with this mine being issued a mining right (Pillay, personal interview, 2018). In retrospect, key individuals understand that the impact of coal mining influenced their constructions of the water quality in general. Their constructions about water quality as influenced by the AMD STT become conflated with the impact of other factors, such as coal mining or sewage spills. The potential effects of coal mining on the water quality therefore become inextricably linked to the effects of AMD STT. The thought processes of key individuals include whether the mine will comply with their water use licence (if issued) or will the mine be involved in illegal discharges into the Blesbokspruit. The mines could possibly attribute their "unlawful" discharges to the STT discharges to "cover up" their pollution or illegal dumping of waste into the

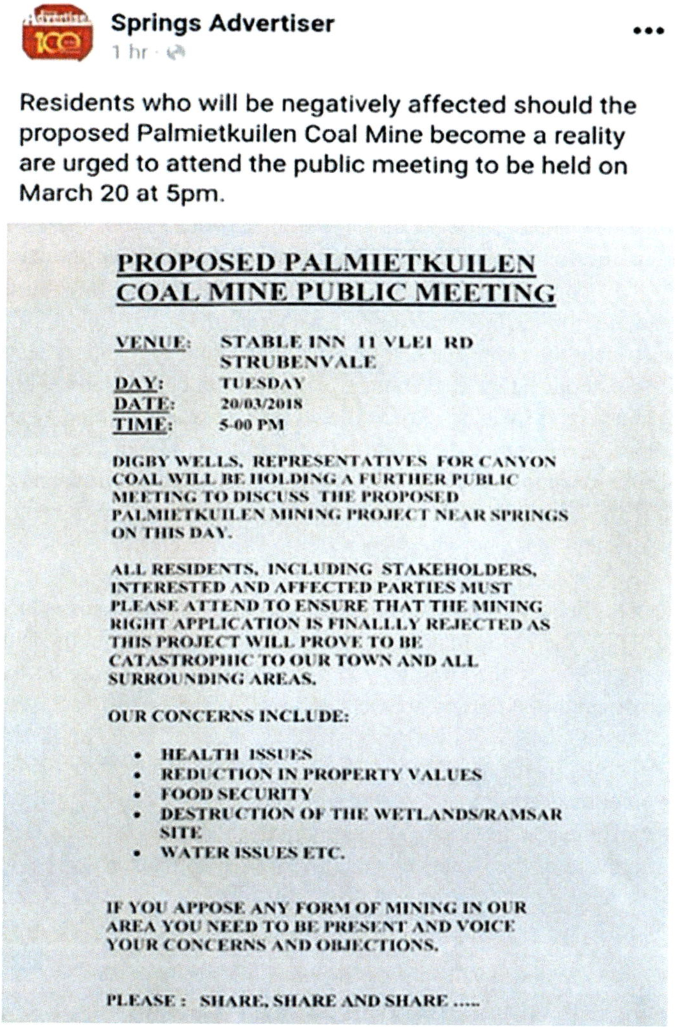

Springs Advertiser

1 hr

Residents who will be negatively affected should the proposed Palmietkuilen Coal Mine become a reality are urged to attend the public meeting to be held on March 20 at 5pm.

PROPOSED PALMIETKUILEN COAL MINE PUBLIC MEETING

VENUE: STABLE INN 11 VLEI RD STRUBENVALE

DAY: TUESDAY

DATE: 20/03/2018

TIME: 5-00 PM

DIGBY WELLS, REPRESENTATIVES FOR CANYON COAL WILL BE HOLDING A FURTHER PUBLIC MEETING TO DISCUSS THE PROPOSED PALMIETKUILEN MINING PROJECT NEAR SPRINGS ON THIS DAY.

ALL RESIDENTS, INCLUDING STAKEHOLDERS, INTERESTED AND AFFECTED PARTIES MUST PLEASE ATTEND TO ENSURE THAT THE MINING RIGHT APPLICATION IS FINALLLY REJECTED AS THIS PROJECT WILL PROVE TO BE CATASTROPHIC TO OUR TOWN AND ALL SURROUNDING AREAS.

OUR CONCERNS INCLUDE:

- HEALTH ISSUES
- REDUCTION IN PROPERTY VALUES
- FOOD SECURITY
- DESTRUCTION OF THE WETLANDS/RAMSAR SITE
- WATER ISSUES ETC.

IF YOU APPOSE ANY FORM OF MINING IN OUR AREA YOU NEED TO BE PRESENT AND VOICE YOUR CONCERNS AND OBJECTIONS.

PLEASE : SHARE, SHARE AND SHARE .....

Fig. 6.5 Public participation meeting in opposition to Palmietkuilen Coal Mine. (*Springs Advertiser* 2018)

Blesbokspruit. Key individuals refer to an activity as illegal if it is not public knowledge. Therefore, it was found that the way individuals understood the impacts of coal mining was historically relative, meaning that individuals formed constructs based on events that occurred in a specific time or from information that was received, instead of questioning these beliefs, their beliefs became patterned based on what they know, such as coal mining and its impacts on water quality.

The DWS showed awareness of the potential impact of coal mining on water resources. They suggested strict criteria being applied in the use of the mine void for the STT for the sludge disposal to ensure that water resources were not further impacted (Tlale, personal interview, 2018). Even though the mining industry has contributed positively to the Blesbokspruit through the constant flow of water, AMD treatment is also contributing to this constant flow. The flow of the water opens up new avenues for business but can impact on tourism, such as the Marievale Bird Sanctuary. Since it is an internationally recognised site, it needs to be protected to regain its Ramsar status (Madden, personal interview, 2018a; Van der Merwe, personal interview, 2018).

This discussion demonstrated that what key individuals from the Springs and Nigel communities know about coal mining and its impact on water quality influenced how they construct the water quality of the Blesbokspruit. Further, this knowledge came from actual awareness of the wrongdoings of coal mines and the government's inability to enforce compliance. This knowledge influences how key individuals feel about future coal mining, based on the current "unlawful" operations. Individuals refer to an activity as illegal if there is no common awareness of it. For instance, the disposal of sludge into the mine void has been described as illegal, because stakeholders were not aware of it, and no environmental impact assessment (EIA) or water use licence exists, which is a legal requirement for such a disposal (Bosman, personal interview, 2018). This enables people to believe an illegal activity is taking place that stems from what policy stipulates is an emergency measure. The DWS has a mandate to protect the water resources through the costly AMD STT. Therefore, their view of the influence of coal mining on the water quality will be similar to the views of other stakeholders, those who have environmental and agricultural interests. To conclude, it is suggested that social issues that arise come from what stakeholders understand about water based on their need for it. Therefore, how it is managed becomes questionable. Their understanding stems from their emotional attachment to water based on their need for it. In this case study, this attachment is fuelled by key individuals' historical experience of mining and their awareness of the issues regarding AMD, as well as the debates over land that should be used for agriculture and that for mining.

6.4 Flows and Volumes of Water in the Blesbokspruit

> *Due to volumes of water in the Blesbokspruit becoming more excessive, the increased water covers the plants, the bird life will reduce, and this impacts on tourism. It's as simple as that.* (Van der Merwe, personal interview, 2018)

The Blesbokspruit no longer flows periodically but continuously throughout the year, due to the various discharges along its watercourse from various industries, as already indicated. The excessive volume of water is beneficial for agriculture, due to the large quantities of water required for its operations. For residents and the tourism sector, the significance of the water volume is twofold: (1) the large volumes of water in combination with the expansion of reeds are beneficial for the birds because they promote diversification of the species and (2) it assists with the dilution of sewage. However, this becomes problematic when flooding, due to reed encroachment, halts the flow of water, affecting their residential properties, and causes a bad smell due to sewage spills. The discharges of water from the AMD treatment plant provide a further dilution effect, which can mitigate the impact of sewage spills.

References to water flows and volumes involve the complete hydrosocial cycle, not only the relationships activated for its access but also those involved in the uses, quality and ways in which its final disposal takes place (Torregrosa et al. 2019). The flow of water of the Blesbokspruit is not natural but co-produced through the mining industry and is maintained through wastewater treatment works and the discharge of neutralised water from the AMD treatment plant. Therefore individuals' knowledge is socially produced. This happens through how they understand reality, in this case through the flow and volumes of water rather than assuming that it simply exists that way. Viewing water as co-produced allows individuals to think about the ways in which social processes shape water and vice versa. Looking at the volumes will reveal the social processes through which water of the Blesbokspruit was formed. The hydrosocial cycle considers how water is manipulated, used and concentrated by social stakeholders, how the struggles for water control (i.e. sewage, reeds, flooding) in this case are expressed in the community. The flows and volumes of water of the Blesbokspruit are impacted by sewage spills, reed encroachment and flooding. The influence that these factors have on individuals' constructions of the water quality is presented in this section. These constructions are based on individuals' realities, which stem from their needs and use of the water.

6.4.1 Sewage Spills

> *While AMD is localised, sewage is not localised; it's all over the Blesbokspruit.* (Van der Merwe, personal interview, 2018)

Key individuals and stakeholders referred to sewage as one of the main causes that impacted on the water quality of the Blesbokspruit. When there is a sewage spill in Bakerton or Strubensvale (residential areas in Springs), the storm water that drains towards the Blesbokspruit is polluted, which reduces the quality of the water. There is a sewage line that crosses the Blesbokspruit, and due to breaches in the wastewater treatment works, "the raw sewage is flowing across the road and into a storm water canal, which flows into the Blesbokspruit" (Matthew Havenga in Blesbokspruit Forum 2019b). This problem further exacerbates the already fragile Blesbokspruit system (Storey, personal interview and site visit, 2018). The poorly maintained wastewater treatment works stem from poor filtration systems (Van Zyl, personal interview, 2016), combined with the human discharge of sewage, which causes further detriment to the water quality of the Blesbokspruit (Humphreys, personal interview, 2018; Labuschagne, personal interview, 2018; Madden, personal interview, 2018a; Storey, personal interview and site visit, 2018). This is informed by cases where raw sewage has seeped into the Blesbokspruit (De Jager, personal interview, 2018b). In addition, sewage spills are also due to illegal miners using sewage water to precipitate the remaining gold. Sewage spills contribute to higher phosphates and nitrates in the water, leading to excessive plant growth and algae.

These sewage spills impact on tourism venues such as Bert Pretorius's Stable Inn Conference and Wedding Venue situated in Springs; "there is black stuff [from sewage] in the water" and this has caused a reduction in the fish life (Pretorius, personal communication and site visit, 2018). Sewage promotes reed growth and eutrophication, which can contribute to a loss of fish life (Govender, personal interview, 2018). Even though there are still fish visible in the water, the quality of water is deteriorating and a further problem is the smell, said Pretorius. Pretorius's venue has a caravan park, meaning that the smell affects his operations. Tourists became aware of these problems and did not want to revisit the venue because of the smell; they complained on social media and were hesitant to camp again even though they find it to be an enjoyable venue. Pretorius's conference facility hosts up to 800 people, and if the sewage impact is left unaddressed, it will reduce his clientele and affect his business, which has been in operation for almost 24 years (Pretorius, personal communication and site visit, 2018).

Van Niekerk (personal interview, 2018), consultant at Karan Beef Feedlot, stressed the fact that solid sewage or untreated discharge from sewage farms or wastewater treatment plants can carry measles, which is a concern for the feedlot industry. Even though Karan Beef Feedlot conducts their own physical surveys of the water quality, its *Escherichia coli* (*E.coli)* assessments indicated whether further assessments needed to be conducted for measles, which was of concern for their business. Back in 2008 Karan Beef Feedlot experienced major problems with high levels of *E.coli* prevalent in the water and possible measles contamination (van Niekerk, personal interview, 2018). Karan Beef took legal action and they won the case against Ekurhuleni Water Care Company (ERWAT). Since then, ERWAT has been working with Karan Beef to ensure better results (van Niekerk, personal interview, 2018). However, *E.coli* was a consistent topic in the Blesbokspruit Forum because of the impact on the water quality and due to concerns about the maintenance of the wastewater treatment plants (Blesbokspruit Forum 2018c).

A common understanding among key individuals was that ERWAT water treatment plants are either overused or just coping and therefore not functioning well (Humphreys, personal interview, 2018; Labuschagne, personal interview, 2018; Pretorius, personal communication and site visit, 2018). Labuschagne, the Ward Councillor for Nigel, revealed his awareness of poor wastewater treatment works that have spilt into the Blesbokspruit and how ERWAT is aware of these implications (Labuschagne, personal interview, 2018). Such implications have been seen in the case of the Balfour wastewater treatment plant that has not been working for over a year, and van Niekerk said he did not foresee any plan that it would be fixed. Some 3 000 houses had been built that would require additional capacity from the Balfour wastewater treatment plant, and concerns were raised about whether capacity requirements would be met to avoid spills into the Blesbokspruit (van Niekerk, personal interview, 2018). Wastewater treatment plants are meant to be monitored by national government, but they have not been monitored consistently, Bedfordview Ward Councillor, Humphreys said. There are water quality guidelines for the Blesbokspruit but in reality, for Humphreys (personal interview, 2018), this is not true; the river is not of that status.

Rand Water, the CoE and the DWS are responsible for conducting water quality tests (De Jager, personal interview, 2018b), and it is often the case that raw sewage is picked up rather than AMD-related problems. Community member, Van der Merwe, has also been involved in conducting water quality samples in the Blesbokspruit for his own research and concluded that when he conducted an analysis of the benthic fauna[6] samples, the species had declined. To him, this suggests that the loss of species is attributed to sewage spills from poorly maintained wastewater treatment plants (Van der Merwe, personal interview, 2018). The large discharges of water from the AMD treatment plant are providing a dilution effect, but sewage promotes reed growth (Govender, personal interview, 2018). The perception is that the large volumes of neutralised water impact on the water quality. However, if the reeds are not well maintained, then this interferes with the free flow of water, creating smells from sewage that impact negatively on business. If the wastewater treatment plants are not well maintained, they have a negative effect on tourism, residents' private properties and agriculture, such as Karan Beef Feedlot, where the cattle are at risk. As a concluding point, based on water sampling and by some businesses conducting their own water samples, key individuals in the agricultural and tourism sectors and community members have developed their knowledge and understanding of how sewage impacts on the water quality. That, in turn, influenced their constructions of how the flow and volume of water have an impact on the water quality. Therefore, a key point was that the volumes of water were not the actual concern, because it was beneficial for the dilution of sewage, but the fact that the flow was negatively affected by the reeds, which is a result of how the reeds are managed (discussed next).

6.4.2 Reed Encroachment

> *My prime reason for worry is the reed encroachment of all the aquatic vegetation.* (Madden, personal interview, 2018a)

Sewage spills combined with reed encroachment further influenced constructions of water quality, stemming from how the water was managed. Water flowing in the Blesbokspruit is both natural and social or human-induced, resulting in changes in the physical landscape. This section explains how the reeds are beneficial for the Blesbokspruit but, if poorly maintained, stagnate the flow of water, which negatively influences individuals' constructions of the water quality.

Before mining became predominant in the area, the Blesbokspruit was merely a little Highveld stream that trickled into the veld during the rainy season. There were no reed beds, meaning the stream flowed completely unhindered through its Highveld grasslands area (Madden, personal communication and site visit, 2018b). Now the water rises to a high level, which is due to the numerous changes (De Jager, personal communication, 2018a) that have led to the expansion of the Blesbokspruit.

[6] This refers to the animals living on, and in, the sediments and large plants of lakes and wetlands (Wetzel and Likens 1991, p. 179).

These changes created excessive water due to years of mining, and now the large discharge of neutralised water from the AMD STT has exacerbated the reed[7] encroachment.

Community member Anne de Jager (personal communication, 2018a) explained how the trees on her land were originally planted by her family when they moved to the area, but now everything has changed due to the increased activities and various uses of the water. Water use has led to neglect and overgrowth of the reed encroachment in the old mining areas of the East Rand, which impacts on the Blesbokspruit (De Jager, personal communication, 2018a). The impact from reed encroachment causes excessive flooding in an already overloaded Blesbokspruit system due to AMD treatment discharge contributing to an influx of water. The reeds, if not well maintained, are known to stagnate the flow of water, which causes a smell from sewage (Pretorius, personal communication and site visit, 2018). Madden (personal interview, 2018a) explained that the most serious concern for him with regard to poor water quality influenced by the excessive reed encroachment of all the aquatic vegetation was the habitat for the birds. This is further exacerbated by the volumes of water and the amounts of nutrients (from various activities, such as AMD treatment, discharges from mining) in the water, including the phosphates, which are poisonous for the birds and the fish (Madden, personal interview, 2018a). The additional chemicals and nutrients in the water exacerbate the reed growth, and as it spreads (Govender, personal interview, 2018; Madden personal communication and site visit, 2018b), it covers the habitat for some species of birds, such as waders.[8] Reed encroachment (see Fig. 6.6) affects the water beds, and there are not many birds left that can adapt from the reeds (Maluleke, GDARD representative, personal interview, 2019).

When the Blesbokspruit Trust was initially formed, there was money available to spray the reeds with herbicide to improve the flow of the river (Van der Merwe, personal interview, 2018). Currently, the Marievale Bird Sanctuary and Grootvaly Educational Centre area are well managed by the GDARD, according to Van der Merwe (personal interview, 2018), but for Madden (personal communication and site visit, 2018b) there is still a fight against the massive reed encroachment. Birders[9] had very negative views of management, such as the CoE and the GDARD. "When the reeds were at its worst" (before 2016) due to poor reed maintenance, leading to bird numbers dropping, this forced community members such as Madden to seek assistance from departments such as the GDARD. However, owing to lack of funding from the CoE and the GDARD to cover wider areas, community members are providing assistance, by managing and maintaining the reeds to protect their businesses and privately owned land (Madden, personal interview, 2018a). For instance, the De Jagers (Philip and Anne) and Bert Pretorius have been doing reed maintenance work by employing people on a full-time basis at their own cost to keep the

---

[7] There are two types of reeds in the Blesbokspruit, namely, (1) *Phragmites* and (2) *Typha capensis* (GDARD representatives, personal interview, 2019).

[8] A wader is a bird species associated with wetlands that live and feed in shallow water habitats (Wildlife Habitat Management Institute 2005).

[9] During interviews a *birder* was referred to by community members as someone who watches birds.

Fig. 6.6 Excessive reed growth in the Blesbokspruit. (Photograph by researcher)

water flow clear and clean (Humphreys, personal interview, 2018; Madden, personal communication and site visit 2018b) (see Figs. 6.7 and 6.8). Reed maintenance creates additional cost implications for these community members (Pretorius, personal communication and site visit, 2018). However, the De Jager's role in reed maintenance occurred long before the establishment of their Riverside Wedding venue (Humphreys, personal interview, 2018). This indicates their emotional attachment to the water because it impacts on their private property.

Through the community's involvement, there has been better management in maintaining the reeds, and through the creation of islands, the flow has improved and has established a decent habitat for birds and the bird life has increased tremendously. However, the reeds cannot be permanently removed, because they have proven to be necessary for the system. It has a valuable impact on removing particular pollutants. Reeds absorb chemicals and other substances from the water into their roots, thereby creating a purification process (Van der Merwe, personal interview, 2018). However, if not maintained, they can be extremely detrimental and contribute to flooding.

According to residents (De Jager, personal communication, 2018a; De Jager, personal interview, 2018b, Madden, personal communication and site visit, 2018b; Maurizi, personal interview, 2018), flooding due to the reed encroachment has been experienced; some residential properties have very little open shoreline left. This means that due to the expansion of the river, the water flows onto the riverbank. Open shoreline is vital and contributes to forming a good habitat that protects birds. Further, it is not only about protecting birds of common species but to build a habitat that protects birds that flock in their millions annually from the northern hemisphere (Madden, personal communication and site visit, 2018b). In Madden's views, unless such protection is provided for such species, there will be a worldwide effect on the birdlife. This has already been experienced in the years when the bird

Fig. 6.7 Reed maintenance at Riverside Country Estate venue. (Photograph by researcher)

Fig. 6.8 Reed maintenance at Stable Inn Conference venue. (Photograph by researcher)

life had declined. About 90% of habitat for birds has been lost due to the reduced shoreline from reed encroachment due to excessive water in the Blesbokspruit (Madden, personal communication and site visit, 2018b). Madden is fond of birding and contributed to the development of Marievale Bird Sanctuary. The protection of the birds is one of his biggest concerns in the Blesbokspruit area.

Despite concerns, in 2018 the number of bird species was at its highest. This proved to be significant, because there have been seasons where the counts have declined due to a combination of pollution, reed encroachment and AMD. The research found that an increase in bird species can be attributed to the AMD STT discharge, because it provides a dilution effect for the sewage. However, the volumes of water can negatively impact on the bird species if the reeds are not well maintained. Selected areas, such as the Marievale Bird Sanctuary, need to be maintained to ensure that the Ramsar Site regains its full status. The GDARD has played a significant effort in this regard (Madden, personal communication and site visit, 2018b). Removing the Blesbokspruit from this record is within the GDARD's institutional interests. Owing to a lack of funding, they focus on maintaining the reeds in specific areas that link to their institutional obligations. Residential owners and individuals from the tourism sector who assist with reed maintenance do so in their own interests. This is based on their emotional attachment to the water that stems from business interests and property values, which is their way of ensuring that the bird life continues to thrive. It was found that the negative perceptions about the discharge of water from AMD STT resulted from how the large volumes of water would negatively contribute to the reed growth and the habitat for the birds. In summary, the constructions of water quality were influenced by poor reed management, which stemmed from the government's lack of funding, rather than it being influenced by the actual physical flow and volumes of the water.

6.4.3 Flooding

> *It doesn't sound right to say, but there is too much water.* (Madden, personal communication and site visit, 2018b)

As already mentioned several times already throughout this book, the Blesbokspruit has more permanent flows with much higher volumes of water than previously. Therefore, the occurrence of flooding has become more regular. Often stated by the community and during the Blesbokspruit Forum meetings was that flooding has been attributed to the water discharged by the AMD STT, by the wastewater treatment plants and by the reed encroachment which blocks the free flow of water. This flow is exacerbated during rainy seasons. This section explains how flooding influences constructions of the water quality of the Blesbokspruit.

A consideration is the regulatory aspects of who is entitled to discharge water into the Blesbokspruit and who should control, manage and decide how the hydrosocial cycle will function (Swyngedouw 2009). Excessive volumes of water causing flooding is a common concern for various key individuals (Madden, personal interview, 2018a; Maurizi, personal interview, 2018; Pretorius, personal communication and site visit, 2018). This concern is explained below.

During a site visit to the Blesbokspruit, community members explained that the Blesbokspruit is vastly different compared to years before, due to increased volumes of water (De Jager, personal interview, 2018b; Madden, personal communication and site visit, 2018b). Evident in the research was that the increased volumes of water in the Blesbokspruit are a result of human-induced events. From what was a small little spruit in the 1940s and 1950 transcended into a series of dams with vastly different dynamics from a wetland (Van der Merwe, personal interview, 2018). "Man's [sic] influence in the early days was absolutely phenomenal as far as changes go, in helping nature and helping nature change itself" (Madden, personal interview, 2018a). However, the increase in the volumes of water has impacts, residents are unable to use their garages because it is partially under water (Madden, personal communication and site visit, 2018b). Private residential homeowners who live on the banks of the Blesbokspruit, which is what attracted them to the area, lose their shoreline. This provides reasoning for concerns with the volumes of water, which stems from "self-interest".

With the increase in water flow, there have been sedimentation issues (Govender, personal interview, 2018). There is now a small flood plain, which is inundated with water, resulting in water runoff on the roads (Van der Merwe, personal interview, 2018). The culverts are unable to handle the flow and the bridges build over the spruit impede the flow of water (Govender, personal interview, 2018; Maurizi, personal interview, 2018), which results in channels being blocked and more ground becoming flooded (De Jager, personal interview, 2018b). The water rises up to 600 mm and 700 mm high, and because it is restricted, it finds the first available open path and flows the easiest way possible, which is up and around the banks where it is unhindered by the flow (Madden, personal communication and site visit, 2018b). If there are swamp conditions, the flow in the system is delayed, and when there is significant rainfall over short periods, the impact of flooding on residential properties along the Blesbokspruit become further threatened through the risk of flooding (Govender, personal interview, 2018). Therefore, the volumes of water create an altered flow that is far too excessive to manage (Madden, personal communication and site visit, 2018b). Flooding impedes on the spatial development of the Blesbokspruit area. It is negatively experienced, especially by the property owners next to the riverbank. Though it is not attributed to the water quality *per se*, the negative impact of flooding overlaps with the release of additional water by AMD STT. The STT is therefore cognitively grouped together with the flooding problems and therefore results in negative social constructions. Instead, the public is blamed for the pollution rather than management failing to improve infrastructure to keep up with growing populations (Pretorius, personal communication and site visit, 2018).

There is more water in the Blesbokspruit than what is needed for the various uses, so the system needs to adapt and adjust to this influx of water (Van der Merwe, personal interview, 2018). Adjusting to the increased water can occur through proper management. However, the CoE does not spend money to clean the culverts; community members assist at their own costs, indicating that "if we didn't assist, we would drown" (De Jager, personal interview, 2018b).

In retrospect, the various concerns about the excessive volumes of water contributing to the flow of the water stemmed from different interests in the water and its uses. These interests differed between government, private properties, agriculture and tourism and therefore had different influences on how key individuals socially constructed the water quality. Govender (personal interview, 2018) narrowed down the problem of excessive water to a water quantity problem rather than a water quality problem, because the water is of better quality now than before it was treated. However, residential property owners refer to the water as being a quantity and a quality problem stemming from poor wastewater treatment works that are further upstream. This resulted in unhappy community members. In contrast, agriculture needs quality and quantity water to conduct its business, for example, Karan Beef Feedlot will have to decide what is more important, having access to water or being more concerned about the poor quality of the water. The water is poor in quality due to the large discharge of neutralised water from the AMD treatment plant, resulting in high sulphate levels in the Blesbokspruit, which can impact on the cattle. However, limited water access will also impact on the cattle. Further, large discharges of neutralised water that contain high sulphate levels exacerbate the growth of reeds, which is already not well maintained by the government departments responsible for doing so. This alters or stagnates the flow of the river, which leads to irregular flooding, which negatively impacts on the bird species. However, the reeds are also valuable for the water and the birds. Therefore, key individuals' constructions were influenced by their interests.

6.5 Technology Used and Processes Followed

> *Without a doubt, such methods are now bearing fruit. We are moving closer to winning the battle against reed encroachment.* (Madden, personal interview, 2018a)

This section looks at how technology used and processes followed to manage the water influence constructions of the water quality. Factors such as pollution-related activities (e.g. reed spraying, wastewater treatment works and the AMD treatment discharges), the technology and processes used in water quality sampling and the technology used and process followed for the STT resulting in discharge of neutralised water and its impact on agriculture are discussed in this section. The purpose is to explain how the perceived impacts of AMD STT are based on what people know about the changes in the physicality of the river due to human-induced events and how the river is managed based on these changes.

6.5.1 "Pollution"-Related Activities

> *We are not going to sit here and fold our arms because of the complaints.* (Joshua, GDARD representative, personal interview, 2019)

Reed spraying, sewage spills from wastewater treatment plants (resulting in high *E.coli* levels) and discharges from the AMD treatment plant influenced social constructions of water quality and attributed to a lack of management and communication. Technology and processes used to manage the reeds, such as herbicide spraying implemented by the GDARD in the Marievale Bird Sanctuary area, led stakeholders to question the impact on the water quality. There have been steady efforts from the GDARD, who have initiated herbicide spraying of the reeds. The purpose of spraying the reeds is to open up the water, and the diversity of bird species is indicative of the improvements made, which meets the GDARDs institutional objectives. Aerial spraying is the method used; the pilot is instructed to spray over the reeds so that the spray does not hit the water. People are employed to physically assist with maintaining the reeds that cannot be reached through aerial spraying (Maluleke, GDARD representative, personal interview, 2019). Specific environmental conditions are adhered to: if the wind is too strong, spraying cannot take place. Water beds also create a challenge and these areas are sprayed first to avoid an overgrowth of reeds. This opens up areas of the Blesbokspruit, ensuring that the water flows more freely. However, aerial reed spraying is very costly (Blesbokspruit Forum 2018d; Joshua, GDARD representative, personal interview, 2019).

Owing to these high costs and limited capacity to manage the reeds in the whole area, the GDARD only sprays the reeds in certain parts of the Blesbokspruit (Blesbokspruit Forum 2018d; Joshua, GDARD representative, personal interview, 2019). The GDARD is guided by a reed management plan, which specifically covers the Marievale Bird Sanctuary area, because its role is to ensure that there is a constant flow of water, and this is where the water best flows through the Blesbokspruit. The GDARD encourages landowners to assist with reed controlling on their properties, which many property owners do. However, this has raised concerns among the public; the GDARD explained that authorisation had to first be issued to those landowners, and they are provided with information on what methods and type of herbicide spray to use (Mashau, GDARD representative, personal interview, 2019). Therefore, clear guidelines are provided (Joshua, GDARD representative, personal interview, 2019). Further, because of the costs, the GDARD requested representatives of the Blesbokspruit Forum and Gauteng Wetland Forum to make suggestions of more suitable methods (Maluleke, GDARD representative, personal interview, 2019). But stated that "we are not going to sit here and fold our arms because of the complaints we have from visitors of the reserve. They come to see birds, and if it [the Blesbokspruit] is full of reeds, then it becomes less attractive" (Joshua, GDARD representative, personal interview, 2019).

Reed management is therefore essential to avoid the infestation, such as bird hides becoming too extensive and unmanageable, and flooding and infrastructural damage, which means if the reeds are not sprayed, additional funds will be required to cover other damages caused (Joshua, GDARD representative, personal interview, 2019). There are only a few bird species that are able to utilise the reeds. Other birds disappear and will not be seen again. The reed spraying contributed to an increase in the bird count, "without a doubt, such methods [stemming from management by the GDARD] are now bearing fruit". For this reason, Madden suggested that "we are moving closer to winning the battle against reed encroachment" (Madden, personal interview, 2018a). However, the GDARD has received mixed views on the reed spraying as it is believed by some to have an impact on the water quality (Jonhasi and Maluleke, GDARD representatives, personal interview, 2019). For instance, Fourie (personal interview, 2019), who is an aquatic ecologist, is of the opinion that there is not enough South African published information on what the impact of the reed spraying is, and there is no consistency in the process. He is of the opinion that people who are given permission to spray may not know how to spray the reeds, and this can harm the life in the water and the water quality itself (Fourie cited in Blesbokspruit Forum 2018d). Even though there were concerns about the effect of the reed spraying on species such as the fish and birds, there was no evidence of this (Jonhasi, GDARD representative, personal interview, 2019).

There is a need to manage and control the ongoing reed encroachment, but individuals were of the opinion that reed spraying affected the water quality in combination with other polluted-related activities. For instance, effluent from industrial activity is dumped close to the AMD treatment plant, in close proximity to the Blesbokspruit, and was not monitored or managed and no fines were issued to the offenders (Labuschagne, personal interview, 2018). There were many concerns about the impact of the ERWAT's wastewater treatment plants on the water quality, because of their plants not functioning adequately. Through the duration of the research, the poorly functioning wastewater treatment plants were said to impact on the water quality. Even though there were improvements in the sampling of *E.coli* levels in the water, it still remained a constant problem. *E.coli* was reported at unacceptable levels at certain sample points (Blesbokspruit Forum 2018b). Owing to evident impacts from wastewater infrastructure leading to high phosphate levels in the Blesbokspruit, associated with leaking sewage pipes resulting in poor water quality (Blesbokspruit Forum 2018b). ERWAT reported that their plants were doing well and would be upgraded, but their wastewater treatment plants proved to have a massive impact on the water quality (Blesbokspruit Forum 2018a). At the beginning of 2018, the Welgedacht and Ancor wastewater treatment plants had complied in terms of *E.coli*. But it is possible that this had more to do with the high rainfall which had a dilution effect on the polluted water rather than improved management of the wastewater treatment plants (Blesbokspruit Forum 2018b). However, in August 2018 *E.coli* was reported at unacceptable levels at certain sample points. To confirm the questions around management of ERWAT's wastewater treatment plants, in September 2020, the water services department at the DWS did not renew their 1-year contract that ended in July 2020, aimed to address the Vaal pollution because "they had not moved at the pace the department desired" (Tshikalange 2020). In addition to these reports and incidents pertaining to sewage, various stakeholders raised concerns about how it could be determined if AMD was flowing into the Blesbokspruit system, since one can detect if sewage is impacting the river (Blesbokspruit Forum 2018c).

While the wastewater treatment works showed to have an impact on the water quality of the Blesbokspruit, the AMD treatment plant discharges showed no signs of impact on the system (Blesbokspruit Forum 2018b). In May 2019 Rand Water reported that there had been an improvement in *E.coli* levels compared with previous years, which created a positive development for the whole system (Blesbokspruit Forum 2019b). During this period, there was heavy rainfall and the AMD treatment plant was not discharging water into the Blesbokspruit due to a fault in the system (Blesbokspruit Forum, 2019b; Madden, personal communication, 2018c; Tlale, personal communication, 2019). Even though the dilution effect from the AMD treatment plant assisted in reducing *E.coli* levels (Havenga, personal communication, 2019a), the rainfall has a better dilution effect, due to the high sulphate levels in the discharged water from the AMD STT. The perception of Govender and Tlale (Sect. 5.6.2) is that the dilution effect of the discharge from the STT is improving the quality of the water, because the poorly maintained wastewater treatment works has impacts worse than that of AMD. However, the water quality samples suggested that *E.coli* levels were high in many periods during 2018 and from mid-2019, and at many sample points, it was reported as unacceptable (Blesbokspruit Forum 2018a, b, c, d, 2019a, b, c, d).

It can be deduced that negative perceptions of the STT were based on poorly functioning wastewater treatment plants and reed spraying. This meant that impacts on the water resulted from processes that were unnatural or human-induced (e.g. technology used and processes followed). This discussion demonstrated that the technology used and processes followed to improve the water quality influenced key individuals' social constructions based on their interest in water and how the water was managed. For instance, key individuals from the DWS and the TCTA form constructions of the water quality based on how they manage the water and how they want the public to view the STT of AMD. Whereas, Maurizi from the CoE and Fourie, an aquatic ecologist, form constructions of water quality based on scientific measurements, such as water quality samples. Individuals from the community construct the water quality based on the visual appeal of the water and their domestic use for water, as well as on environmental interests, such as the diversity of birds. Therefore, for these various individuals, reed spraying, for instance, will be beneficial, because it improves the flow of the water, which influences their constructions of the water quality. As a concluding point, the AMD STT discharges somewhat contributed to the dilution effect of sewage and influenced individuals' constructions of the water quality.

6.5.2 *Water Quality Sampling*

> *How would one account for, and include the impacts on, water quality coming from non-source points, where sampling does not take place.* (Tlale, personal interview, 2018)

A crucial part of the Blesbokspruit Forum meetings was for the presentation of water quality reports by entities who conduct various samples of the water, namely, the DWS, the CoE, ERWAT, the Lesedi Local Municipality and Rand Water (Blesbokspruit Forum 2018a, b, c, d). The purpose of these reports was to indicate the quality of the water and determine whether the water was deteriorating and if so, what the causes were and how it can be better managed. Based on the known uses of water of the Blesbokspruit, samples are taken to measure the quality of water and determine any impacts. The geographical location where the sample was taken is referred to. It is therefore possible to see what the actual issue is and who was responsible for the pollution and whether authorisation had been provided for that activity. However, water quality sampling proved to have its own challenges, such as ensuring the safety of the officials at the sampling sites and the flow of water compromising the accuracy of results. This led to stakeholders' questioning the credibility and legitimacy of the process, resulting in a contributing factor influencing how water quality is socially constructed. Water quality sampling influenced constructions because scientific measurements/evidence of the water quality should be more credible than mere visual observations or reports, but complications with the sampling process can undermine this credibility and contradictory sample results can have the same negative effect.

DWS representatives expressed their concerns about safety when conducting water quality samples. This was due to the presence of illegal miners. Pillay (personal interview, 2018) illustrated that "Zama Zamas can be rough and tough, and dangerous". Concerns arose because there was a lack of security at the sample sites (Blesbokspruit Forum 2018a). Govender (personal interview, 2018) suggested that more often than not the safety issue was driven by a perception of insecurity; "if I go there I will get attacked". He was not aware of such a situation but supported the perception because he would not want an officer to be at risk while taking samples, unless they had an armed escort. The Ekurhuleni Metro Police Department has escorted CoE officials to obtain their water samples (Maurizi, personal interview, 2018). Maurizi suggested that crime in general was a problem, and the samples were taken in the middle of the river, implying that the sample sites were isolated. Despite these concerns, Maurizi and Govender are both of the opinion that it is unacceptable for the sample not to be taken, because there are many other points where samples can be taken that are representative of how the Blesbokspruit is performing (Govender, personal interview, 2018).

In addition to the safety concerns were difficulties with acquiring samples due to the flow of water. A consistent flow of the water is required in order for the water to be sampled at specific points of the Blesbokspruit (Blesbokspruit Forum 2018a). A CoE representative (Blesbokspruit Forum 2018a) explained that not all sample points are monitored, because sampling can only be done once a month and no later than 10:00 in the morning to get an accurate sample of the water, before the daily discharges take place. There is no ideal method of sampling the water, but to look at the quality of water over a year and to identify changes in colour of the water, even though this is not the best option, it provides a clearer picture of the water quality (Rand Water representative in Blesbokspruit Forum 2018a).

Concerns over safety and the flow of the water led to concerns about the accuracy of testing, as expressed by members of the Blesbokspruit Forum. Sometimes unforeseen challenges arise such as a community protest, leading to the whole area being closed and people's lives cannot be placed at risk to obtain a sample (Maurizi, personal interview, 2018). If sampling occurred sporadically, then other releases and spillage points would not be recorded. The longer the frequency between sampling, the more difficult it is to know what is taking place on the days where sampling is not conducted, because something can go wrong (Blesbokspruit Forum 2018a). At certain sample points there is only storm water, so a sample cannot be taken if there is no rain to ensure a flow in the water (Maurizi, personal interview, 2018). Maurizi (personal interview, 2018) explained that water always flows to the lowest point and sometimes creating a blockage. Even though sewer lines extend for hundreds of kilometres and there are many pump stations that are monitored, no one is available on a daily basis to clean blockages. However, if there is a real cause for concern linking directly to sewage pollution, then the CoE manages it immediately (Maurizi, personal interview, 2018).

But a massive concern was how one would account for, and include, the impacts on water quality coming from non-source points, where sampling does not take place, and those who do not have licences to discharge waste, if stakeholders referred to testing as inaccurate (Tlale, personal interview, 2018). Further, illegal miners can contribute to polluting the water by blocking sewage pipes, making it difficult to gain access to that area of water to take a sample, because the blockage can happen in the middle of nowhere (Maurizi, personal interview, 2018). Further, government officials are intimidated by illegal miners, creating a challenge in reducing this source of pollution. Such pollution incidents could impact on the AMD treatment process and create negative perceptions of the water quality among stakeholders, which is a concern for the TCTA. Tlale (personal interview, 2018) indicated that accuracy of testing had nothing to do with the TCTA and its operations with the AMD STT. The TCTA has a storm water management plan to account for such impacts, which was necessary considering the costs, monitoring and rehabilitation that go into the AMD STT (Tlale, personal interview, 2018). This is confirmed by Rand Water reports (Blesbokspruit Forum 2017) that there did not appear to be an indication of the AMD discharge in the sampling and monitoring results since mid-2016, when the treatment plant became operational.

During Blesbokspruit Forum meetings, members debated and discussed the water quality results from the different entities. Early in 2018 there was an increase in sulphate levels between the Blesbokspruit and Marievale Bird Sanctuary. This was linked to AMD, but there was no clear indication what the increased sulphate levels were attributed to, and by the time it reached the Blesbokspruit, there was a dilution effect. Liefferink (Blesbokspruit Forum 2018a) explained that dust flowing into storm water can increase sulphate levels. At the end of 2018 sulphate was reported at unacceptable levels on the water quality reports, due to high content at Heidelberg, where Karan Beef is situated; this is an indicator of AMD. This was the first time that levels were high since the STT discharge started (Blesbokspruit Forum 2018d). However, during mid-September 2018 until January 2019, the AMD treatment plant was not operating due to a fault in the system (Madden, personal communication, 2018c; Tlale, personal communication, 2019). This meant that other sources of pollution, combined with the high sulphate levels already in the water, exacerbated the effect. In August 2018 the Balfour wastewater treatment plant was reported as non-compliant in terms of sulphate, and it was the only sulphate non-compliant point in the catchment (Blesbokspruit Forum 2018c). However, Rand Water samples and that of other departments differed at times, and it was suggested that the DWS water quality reports include a column indicating the sample date and quarter to make it easier to understand (Blesbokspruit Forum 2018c). This was to improve the accuracy of the sample. ERWAT's water quality reports often indicated that there was improvement with regard to *E.coli* downstream of the catchment where most spillages occurred but mostly reported positive samples, which somewhat did not match the actual water quality reports of other departments or reported sewage spillages into the Blesbokspruit. This could imply inaccuracy of testing or unknown sources of sewage spills.

This discussion implies that the perceptions of the discharge of neutralised water from the AMD STT were linked to the process followed to conduct water quality sampling and reporting. This was evident based on the opinion of some individuals, who suggested that the results of the samples could not be indicative of the actual quality of the water, because of the concerns raised with accuracy and consistency in sampling. This meant that, although the purpose of sampling is to provide scientific evidence, which is meant to be credible, this section indicated that the different results and uncertainties in water samples from various sources of pollution influenced the constructions about the water quality.

6.5.3 Neutralised Water and Agriculture

Different crops can grow in different soils conditions. (Nell, personal interview, 2016)

There are agricultural businesses situated directly along the Blesbokspruit. These businesses have centre pivots where their farms are situated that are used to pump water out of the Blesbokspruit for irrigation. The amount of water used per year for irrigational purposes depends on the water use licence granted to farms (Agricultural sector representative, personal interview, 2018). Stakeholders raised concerns about what possible impacts the discharge of the neutralised water into the Blesbokspruit would have on agriculture. The neutralised water is the result of the AMD STT process based on the technology used to partially treat the water. The concerns raised were due to the high sulphate content remaining in the discharged water. Agriculture is dependent on water for irrigation. The concern of some agricultural interest groups related to the consequences that high sulphate levels would have on the crops and the cattle. Based on this, the discharge of neutralised water influenced how individuals (e.g. from the agricultural sector, community members, local government, researchers and the AMD project team) socially construct the water quality.

Concerns about water quality were mainly for downstream water users (as already explained earlier in this book), where Karan Beef Feedlot is situated. Van Niekerk, who is the Blesbokspruit Forum representative for Karan Beef, reported that the water quality has not reached a critical point as yet, but it was very close

(Blesbokspruit Forum 2018b). He conducts statistical analysis of the probable impact that the high sulphate content can have on the water and suggested that it would be a concern if it continued to be discharged at a high rate because it could affect the growth rate of cattle and the cattle feed conversion[10] (Blesbokspruit Forum 2019b). "Water is the lifeline of a feedlot and that is why we have to protect it every day to make sure we don't get sudden bursts of dirty water coming through" (van Niekerk, personal interview, 2018). Each cow needs 40–50 ℓ of water per day. Even though "the water can be a little dirty, cattle can drink it, because they do not have to drink water that is of potable standards" (van Niekerk, personal interview, 2018). However, higher levels of sulphate in the water can be problematic and must therefore be monitored (van Niekerk, personal interview, 2018). Van Niekerk reported in 2018 and 2019 that water quality for the cattle had been very good and sulphate levels had been at their lowest since 2014 (Blesbokspruit Forum 2018b, 2019b). Based on this, van Niekerk does not think that there are increased sulphate levels, but sulphates tend to saturate water and the slightest change in movement of water, through rain or runoff, will create a more regular problem (Blesbokspruit Forum 2018c; van Niekerk, personal interview, 2018), which can eventually change the soil structure, for instance (Fourie, personal interview, 2019; Van der Merwe, personal interview, 2018).

However, according to Fourie, a good farmer knows to look at the soil salinity and the nutrient content and fertilise accordingly. If there are any differences in the water quality experienced, then farmers would not clean the water; they would simply adjust their soils (Agricultural sector representative, personal interview, 2018). If there was a drastic change in the production of the centre pivots, then the farmer would look at the water quality and the effect it has on the soil. There are benchmarks (e.g. aluminium in the soil is disastrous for crops and restrict the access of nutrients into the plant) to determine if any water quality issues have surfaced (Agricultural sector representative, personal interview, 2018). Agricultural businesses test the soil every year to ensure that there is a correct pH balance (Agricultural sector representative, personal interview, 2018).

Nell (personal interview, 2016), a researcher at the Agricultural Research Council (ARC), explained that different crops can grow in different soil conditions. Each type of crop has a fresh salt value, and only once this value has been reached a decline in the yield is visible. People thought this would be disastrous, but Nell suggested that after monitoring the Blesbokspruit for over 15 years, he found the soil to be in a better condition. For example, with the same water quality, cotton is salt-tolerant but beans are less tolerant. "People think if you cannot grow beans or carrots it's not suitable for agriculture, but you do find certain crops that are salt-tolerant" (Nell, personal interview, 2016). Maize, however, depends on different levels of water quality and salt content, and if the level is high, then it can affect humans. If water from the Blesbokspruit is used for vegetable production, then it would impact

[10] *Cattle conversion feed* means "the amount of feed an animal consumes as compared to the amount of body weight gained, expressed as a ratio" (Reuter 2009).

on food safety (Nell, personal interview, 2016). Crops such as soya bean and the small white bean do not retain a lot of moisture, meaning there would be traces of heavy metals. However, watermelons, pumpkins, lettuce and spinach retain water, and if this water is of poor quality, they would not be fit for human consumption (Agricultural sector representative, personal interview, 2018). This means that the sulphates provide nutrients in the water that are beneficial for some forms of agriculture and farmers do not complain because the water is high in nutrients (De Jager, personal interview, 2018b).

The neutralised water from the AMD STT can therefore be used for agriculture (Nell, personal interview, 2016). The eastern basin has not had loss of agricultural land due to AMD spillages as in the western basin (Naidoo, personal interview, 2018), where "poisonous stuff [was] coming out" (Agricultural sector representative, personal interview, 2018). Some farmers were known to sue mining companies because of this (Nell, personal interview, 2016), and it is easy for them to get money this way, because acidic water affects the pivots. Farmers who are situated in coal mining areas suffer the worst impacts on agriculture; spills can have a lasting impact (Nell, personal interview, 2016). Therefore, the concerns about the impact of future coal mining and its effects on agricultural land are far worse (discussed in Sect. 6.3). Based on farmers' interests in water use and how they understand the impact of the neutralised water influenced their construction of the water quality. These include farmers who probably never thought that the day would come when they would not be able to plant their crops and irrigate using river water because of the problem with the water quality (Agricultural sector representative, personal interview, 2018). These concerns arise because the quality of water needed for agriculture is different for certain commercial farmers and subsistence farmers (Fourie, personal interview, 2019). Subsistence farmers along the Blesbokspruit use the water to irrigate their plants and high sulphate levels can cause bacterial contamination in the plants or vegetables. This would impact on a subsistence farmer such as community member, Stewart (personal communication, 2019), whose sole income is from the crops (vegetables) he sells to the local towns. He is solely dependent on borehole water.

If there are high levels of sulphate in the water, even though it can be used for agriculture, the DWS and Rand Water do not approve of this quality because this water cannot be used for domestic use or subsistence farming (Nell, personal interview, 2016). Although Govender (personal interview, 2018) suggested that even though the sulphate levels were high, there was a dilution effect, which made the water safe for agriculture; the wetland had its own natural cleaning process. However, Stewart (personal communication, 2019) said that he still had the same concerns about the neutralised water as before the STT because these concerns were not addressed during the EIA public participation process. AMD is therefore not the only problem for surface water quality, but that of groundwater availability, to complicate the matter further. Potential future mining in the area will impact on the water quality due to increased sulphate levels from their discharges (van Niekerk, personal interview, 2018). A business as large as Karan Beef has to conduct risk assessments to ensure the well-being of its livestock, because it has a large number of cattle (Blesbokspruit Forum 2018b). Even though Karan Beef's meat quality is very good (De Jager, personal interview, 2018b), if the water quality does not improve, it will have an impact on the cattle in the long term (De Jager, personal interview, 2018b; Humphreys, personal interview, 2018; Mashau, GDARD representative, personal interview, 2019; Van der Merwe, personal interview, 2018). While the high sulphate levels are alarming and pose a possible threat to the cattle in terms of food safety, lack of water availability in general poses a bigger threat to food security.

Different perceptions were evident between individuals in the agricultural sector about the impact of neutralised water from the STT. These perceptions differed based on what types of crops were irrigated with this water, which were linked to business interests. Water quantity is as important for agriculture as water quality. The large discharges of neutralised water are therefore beneficial for agriculture but dependent on what types of crops are produced with this water.

6.6 Information and Communication

> *Why is the public kept in the dark? Why is only a handful of public people made aware and others not?* (Klip River Forum 2018)

This section illustrates how information and communication are factors that influence social constructions of water quality. Knowledge about water comes from somewhere, it is local, and understanding how knowledge travels will enable one to understand why it is viewed that way. Social constructionism argues that the ways in which individuals commonly understand the world and the categories and concepts used are historically and culturally specific (Burr 2015, p. 2). There are many ways of looking at the same thing, influenced by access to information and communication.

The AMD project was declared a government waterworks that had to be constructed under emergency circumstances and was therefore exempted from environmental authorisation (Govender, personal interview, 2018). Arguably one of the most influential factors of information and communication for this discussion is the non-use of the approved EIA. Though it was meant to be an important step in the planning phase of the STT, it also served as an important medium of communication and distribution of information. In an ideal situation, if the AMD project was complied with correctly, the STT would have involved two approved EIAs: one for the AMD treatment plant and discharge of neutralised water and the other for the eventual sludge disposal site – the mine void. Then, communication in various ways could have been possible: the way in which responsible government departments presented their planning, the affected parties presenting their views or assessments of the planned processes, sharing of information and hearing the views of the different stakeholders. If the AMD project followed this ideal process, then it could have had a different impact on perceptions and social constructions before the STT was even implemented. However, the way in which the project unfolded and was implemented led to the negative perceptions developed during the planning phase and also created negative constructions of the water quality once it was implemented.

Since the STT of AMD was implemented under emergency circumstances, information about it was not shared widely and publicly. This was evident in stakeholders' perceptions after the STT had been implemented, which centred on the purpose of the EIA process (discussed in Sect. 5.6.3). These perceptions stemmed from the mine void being used as the sludge disposal site, which was different from the proposed site assessed in the EIA and the fact that the construction of the AMD treatment plant was exempted from an EIA. This already implied that the information about the STT was not public knowledge, as no EIA public participation process had taken place. This lack of information and communication influenced how stakeholders and key individuals socially constructed the water quality of the Blesbokspruit based on the unknown impacts of AMD STT. The purpose of an EIA public participation process is to ensure the transfer of information to the public on the proposed project, then for the public to communicate their concerns and suggestions and for the project team to respond. Therefore, the expectation from the public would have been for ongoing communication regarding the STT. However, because the new sludge disposal site for the STT was implemented without the public's knowledge, it was viewed as serving the interests of the AMD project team and not the community.

Not only was there limited access to information, but existing concerns were not clarified. During the EIA public meetings, there were knowledgeable stakeholders present who knew about relevant policies and processes and who asked valid questions. They were also able to question the suggestions and responses of the project team. However, the questions and suggestions from these knowledgeable stakeholders were not fully addressed (Maurizi, personal interview, 2018), and this could have created concerns for other community members and other stakeholders who were present at the meeting. Further, there could be a very good speaker, who is also emotional, but everyone listens to that person rather than the scientist (Van der Merwe, personal interview, 2018). Bosman[11] (personal interview, 2018) critiqued the apparent thinking of the government that the community was unaware, while she felt that the community was knowledgeable and actually asked the right questions but was fed wrong information. This resulted in the Springs community being kept in the dark, suggested Ward Councillor Labuschagne (personal interview, 2018) (Fig. 6.9). Community members often informed Labuschagne and provided him with direct information. This implied that he was also part of the public that was kept in the dark. Therefore, the transfer of information took place in various ways. Labuschagne shares the information he receives with the rest of the community in his ward by reporting the information to local newspapers. Ward Councillor Humphreys (personal interview, 2018) (Fig. 6.9) communicates to the community through social media and the impact is evident in the fact that her followers are

[11] C. Bosman conducts training in water use licence applications and all relevant policies relating to water use.

Fig. 6.9 Ward Councillors Wollaston Labuschagne (left) and Jill Humphreys (right). (Photograph by researcher)

growing. Van der Merwe (personal interview, 2018) holds meetings with the community to keep them informed and allows them to ask questions. This demonstrated how information is transferred through many other means by influential individuals, which should, instead, be the government's role. The government does not pay the public much attention, and the community is always the last to be informed, which leads to incorrect information spreading and ongoing unhappiness (Naidoo, personal interview, 2018; Van der Merwe, personal interview, 2018). Govender (personal interview, 2018) agreed that there was probably a lack of communication and explained that the DWS hoped to rectify this through the Blesbokspruit Forum and media communication.

The different means of communication, such as during the EIA process and through the Blesbokspruit Forum, could have created a shared understanding of the negative impacts of the water quality that stemmed from the STT. Communication is the most essential means of recreating reality, and because there was no clarity on issues that stakeholders raised, existing concerns about the treatment started to develop and were shared among stakeholders. These negative perceptions of the water quality were produced before, and remained after, the AMD STT had been implemented. This created socially produced ideas of the negative impacts of the neutralised water that is discharged into the Blesbokspruit and the potential risk of the sludge being disposed into the mine void. These socially produced ideas stemmed from the transfer of the incorrect information provided during the EIA public participation process, such as the mine void being unfeasible, but then became the implemented sludge disposal site without participation by the public or an explanation for the choice of site.

This incorrect transfer of information during the EIA process meant that stakeholders formed mixed views about Digby Wells Environmental as the EIA consultant and their ability to listen to the community's concerns. Ironically, when the treatment plant launched, Digby Wells was not aware that the sludge was being disposed of in the mine void and not the site for which they had conducted the EIA (Pillay, personal interview, 2018). Owing to their lack of information, Digby Wells was of the opinion that water quality would start deteriorating, because sludge disposed of into the mine void would eventually rise, as no one knew what was happening underground (Pillay, personal interview, 2018). Thus, they were placed in the same position in which the stakeholders were during the public participation process where participants felt that they had limited access to information. More so, communities felt that some of the information in the EIA reports was incorrect, such as the number of boreholes identified (Digby Wells Environmental 2014). This created a perception that findings in the report were inaccurate. Such inconsistencies created an ongoing disbelief and lack of trust in the process, influencing the constructions of the water quality. Inaccurate data provided during the EIA process led to questioning of the legitimacy of the actual STT (De Jager, personal interview, 2018b).

Bosman (personal interview, 2018) felt that when AMD first became an issue, a strategic plan could have been drafted in which scientists from various disciplines should have been involved. However, because this was not done, the options from a legal, economic, social and environmental perspective were never properly addressed, and this is what led to a site-by-site process (Bosman, personal interview, 2018). Private sector ideas are not used for solutions and stakeholders do not understand why (Barker, personal interview, 2018). This meant the fact that the treatment was implemented as an "emergency" implied that people would not have had the opportunity to ask new questions or be informed correctly of the new processes. Bosman (personal interview, 2018) suggested that referring to the STT as an emergency prevented exact and proper scrutiny and could have been deliberate in order to prevent the right questions from being asked that the project team could not answer. In either case, the public's ideas of the STT were influenced by the communication (or lack thereof) and information provided, and this is how individuals socially construct the water quality. It was based on what they heard or did not hear.

Critically significant information was often not shared with the public. One such example is the following: Any environmental risk that stems from the AMD STT will not lead to decanting (Tlale, personal interview, 2018), because there is a monitoring process with the sole purpose of maintaining the water level below the environmental critical level (ECL). However, this was not public knowledge. Such information is shared and explained when stakeholders ask about the potential impacts during meetings of the Blesbokspruit Forum, for instance. Another example refers to the impact that the STT has on the water quality, which is actually only confirmed through the water quality sampling. This has also proven to be untrustworthy, based on the concerns and issues of safety of officials who have to take the samples and this questioned the accuracy of the sampling.

Limitations on public access to the sampling information have a direct effect on perceptions and social constructions. In 2018, for example, a presentation to provide an update on the treatment in the three basins was provided during the Klip River Forum. When questions were asked, a DWS representative indicated that specifics could not be divulged (Klip River Forum 2018). Further, the TCTA said that it can only disclose its water quality results once permission from the DWS was obtained. The forums are public platforms for communication and to share the water quality results of the Blesbokspruit, leading a representative of the forum to ask "why is the public kept in the dark? Why is only a handful of public people made aware and others not?" (Klip River Forum 2018). Even though Govender (personal interview, 2018) of the DWS stated that communication should be enhanced through the forums, when forum representatives tried to gain clarity, the response from DWS officials remained unclear. Another example of how communication complicated the perceptions and constructions is when stakeholders were informed during forums that high salinity levels come from mine discharges and that it was difficult to attribute the salinity changes to a specific source, yet new mining rights were issued in an area undergoing excessive rehabilitation with limited compliance monitoring measures (Blesbokspruit Forum 2018c). If officials cannot answer questions and provide information, then the public cannot understand or be accepting of the process. The forum meetings take place on a quarterly basis and this is the only platform where clarity can be obtained and communication can take place. When stakeholders tried to obtain information, it was not always readily available. This means that these communication deficiencies influenced individuals' constructions of the water quality.

There are mixed views about the quality of water because so much is happening in the Blesbokspruit catchment and the plan for the long-term treatment (LTT) is not well communicated for stakeholders to change their perception of the STT. They are not convinced it is a measure only for the "short term". Therefore, stakeholders want to know how the high sulphate levels in the water will be controlled, when will the LTT begin and if the DWS does not have the funds, who is going to pay for it (Barker in Blesbokspruit Forum 2018b). The private sector and investors do not know what is happening and are unhappy about not knowing about the LTT to address water security, because it has business implications.

It is therefore evident that perceptions of AMD STT could have been less negative if information and communication were more transparent. It is illustrated in the following two examples from key individuals: The first example is that of Fourie who suggested that there was a vast difference between AMD and AMD STT. Everyone was concerned about the treatment, but they should remember what the deteriorated wetland looked like; it was very bad. Despite the sewage spills, there were still beautiful flowers growing and bird numbers were increasing (Fourie, personal interview, 2019). When the treatment plant stopped pumping in December 2018, the Blesbokspruit was running low. This means that the system is dependent on the water from the STT (Blesbokspruit Forum Google group 2018). Fourie asserts that the system keeps functioning despite the impacts. Therefore, the discharged water from the AMD STT is actually advancing the system. The second example is that of Van der Merwe who is more aware of the STT, because he is in contact with the AMD project team and he asks questions for clarity, which reduces his concerns because it is a short-term arrangement. Van der Merwe and Naidoo, both Blesbokspruit Trust members, do not think there will be an impact, because the sludge is being disposed so deep down that it will take a very long time for the pH of the water to change (Van der Merwe, personal interview, 2018). This is an indication of more positive views of the STT, based on individuals' having insight. This in turn will lead to less negative perceptions on the quality of the water. However, for those who do not have this understanding, their perceptions of water quality remain more negative.

The Blesbokspruit Forum is a useful platform and plays a significant role when it comes to information and communication (Maurizi, personal interview, 2018) and can therefore be used fully for this purpose. Further, to ensure better communication and effective information sharing among group members, a Google group was created by a representative of Rand Water for members of the Blesbokspruit Forum. This platform also allows the water quality reports and other important documents to be shared before forum meetings take place (Blesbokspruit Forum 2018c).

A significant finding was that if access to information and communication was more transparent, it would lead to less negative constructions of the water quality and therefore individuals' negative perceptions towards the AMD STT. Participation is crucial to this process. Stakeholders' constructions of the water quality come from what they hear from particular interest groups. How they interpret what they hear depends on whom they trust and gain access to information from, and this depends on who forms part of the Blesbokspruit Forum, for instance, and whose views of these particular interest groups they trust.

6.7 Vested Interests

> *Different people have different interests [with water] and therefore it brings different concerns to each.* (Naidoo, personal interview, 2018)

This section explains how facts of water do not speak for themselves but in reality what counts as the truth about water varies depending on the perspective of who is talking about it, their purpose and what the individual interests are. Interests can be institutional or personal, material, reputational or status interests. Many social science theories and approaches (such as public choice theory or pluralism) use interests as their focus of attention. They explain relationships between positions (perceived or real) and decision-making to protect or strengthen these positions. Social constructions articulate the consciousness of these positions and how they can be affected by decisions or changes in the social conditions, such as the use of water.

Individuals provide different reasons when describing the same water quality. These reasons do not reveal or hide the truths of water quality but, instead, create their own individual truths. Whose reason is accepted as being truthful is a question of social struggle, interest and power politics. "Different communities have different agendas, different people have different interests and therefore it brings different concerns to each" (Naidoo, personal interview, 2018). The aim of this section is not to identify who is right or wrong, for instance, between the mining company, the community, the activist or the scientist, but to show that realities are created by individuals in specific historical and material practices.

This was illustrated in the array of vested interests that were visible in the Palmietkuilen Coal Mining project and the EIA. Pillay (personal interview, 2018) observed that the "usual suspects" (i.e. environmental activists and the known community members in Springs) were challenging this coal mining project. However, there are community members – from, for example, the informal settlements or townships – who will support the project because of the potential opportunities for employment. Further, environmental consultants have to look at the environmental safety components first and present what they have assessed. This may differ from community activists who focus on the interests of the community first and might reject the assessment of consultants whom they might accuse of being paid-off to favour the project of the mining company whom contract them (Pillay, personal interview, 2018). Residential owners might think that the mining company arrange for the people from the informal sector to attend the EIA (called "renting a crowd") to support the project as they have an interest in the promised jobs linked to the mining project (Agricultural sector representative, personal interview, 2018; Tlale, personal interview, 2018). The interest of residents from the informal settlement might be that they were promised employment (Agricultural sector representative, personal interview, 2018; Labuschagne, personal interview, 2018). A range of interests is then possible in the mining project, from potential employment opportunities to protecting the environment and maintaining current livelihoods and values of residential properties.

These various divergent interests can lead to conflict between those seeking to ensure that their interests are served. More so, the most influential ones are the ones who are heard. The DMR will promote the mine; they would issue a mining right because their purpose and interest are with economic growth, but the protection of the environment is excluded in pursuing these interests (Maurizi, personal interview, 2018). The most influential person's view is considered, in this case, the national government, which has the authority to make the decision (discussed in Chap. 7). The various interests depend on the individual's needs, for business (i.e. environmental consultants, agriculture, mining and tourism), livelihoods (i.e. residential property owners) or employment (i.e. informal settlements). The argument from those opposing the mining operation was that the employment would be temporary just as the mine would be, and the impact on the environment would be long-lasting. However, it is difficult to convince a poor person to think of tomorrow, that is, the environmental impact. She/he would rather take that job for 1 year to sustain her/himself for the time being (Agricultural sector representative, personal interview, 2018). This means that "the desperation out there completely influences people choices" (Humphreys, personal interview, 2018).

The influence of authoritative parties to make decisions is further explained with the STT. The DWS and the TCTA are aware that the discharged water from the AMD STT is not to a specification that the law (i.e. National Water Act (NWA) and National Environmental Management Act (NEMA)) states it should be, indicating awareness that the water is not the ideal quality for the environment (Govender, personal interview, 2018; Tlale, personal interview, 2018). Even though they show awareness that the water is regarded as polluted water, because the water is now partially treated, the government should not adopt a mindset that it is acceptable and sit back because it has done something. This water will only be suitable for a limited period (Van der Merwe, personal interview, 2018). This is a case of the government (through the DWS) focusing on what is in its interest and within its jurisdiction and what is feasible at the given time to pursue its interests, addressing issues for which it is responsible.

Maurizi (personal interview, 2018) acknowledged that government has dragged its feet when it came to deciding and implementing the AMD STT but suggested that given the nature of the Blesbokspruit, the AMD rehabilitation had to also be understood from a financial perspective (i.e. what funds are available). Her view in this regard comes from the interest of being a Water Quality Manager at the CoE. She explained that while people are living around the Blesbokspruit, the impacts would continue and all stakeholders involved have to be realistic about what was needed and what could be done, even if it was not the ideal situation for the environment (Maurizi, personal interview, 2018). The idea is that in an environment there are certain complexities that can be managed and other aspects cannot. If new complexities are created, the decision will be made as to whether or not these complexities can be managed and, if they cannot be managed, then the decision is made to focus on what is manageable (Barker, personal interview, 2018). There is a vested interest that comes with water management, "the intention is water for business, not the environment" (Barker, personal interview, 2018). This implies that water for business is favoured over that of the environment.

However, the government has to consider the various uses other than business use of this water and not only look at reducing the heavy metals and high sulphates. They have to consider other uses such as domestic use, where the water still cannot be used even if these chemicals are reduced (Maurizi, personal interview, 2018). Therefore, individuals felt that indicating that the water was of a better standard did not mean that the quality was acceptable to all its users. This is the value and relevance in conducting an EIA to assess such complexities, and it links to the quality that all users require. Determining what quality of water one is looking for is determined by individuals' use for the water (and that is determined by their interests). It is therefore not possible to use a variable to measure water quality that is specific to all industries (as discussed in Sect. 5.2). The examples to follow indicate key individuals' and stakeholders' acceptance levels of the water quality in the context of the AMD STT, indicating the various interests with the water.

Some commercial agricultural businesses are still able to use the neutralised water produced by the STT, depending on the crops they produce, but subsistence farming cannot, because of the quality of water required to grow vegetables.

"Ironically, the farmers did not want the water to be taken out of the catchment, because downstream they are going to suffer once there's a reduction. So they were in support of water continuing to flow in, but from an environmental ecological perspective there was a concern of neutralised water discharging and its impacts" (Pillay, personal interview, 2018).

This is why some individuals from the agricultural sector indicated that they had not had problems with the water quality that they used for crops (Agricultural sector representative, personal interview, 2018). Even though the high salinity levels have been a concern for Karan Beef and their cattle, for instance (van Niekerk, personal interview, 2018), it does not disadvantage other commercial farming operations because both the quantity and quality (high nutrients) are necessary for such a business (Nell, personal interview, 2016). Impala Platinum was also interested in using the neutralised water from the STT, "the water purity level is one step away from potable water and therefore it is definitely of better quality" (Naidoo, personal interview, 2018). Impala found this water quality to be acceptable to use for their business operations. Community members Van Blerk (personal communication, 2019) and Ferreira (personal communication, 2019) have not experienced any negative impacts on the water quality as yet from the STT. However, they suggested that even though the treatment plant was a nice set-up, the concerns raised during the EIA process were not addressed, and they were still fearful about their borehole water, which is the only source of water for some.

Holden (personal communication, 2019) from the TCTA explained that the TCTA and the DWS had drafted a Stakeholder Engagement Strategy for a STT in 2011, long before the EIA process had begun, and it did not include a discussion on the sludge disposal site. The whole process involved a "battling problematic perception", he said. The responses were based on misguided belief, and for others it was due to deliberate misinformation put out by those whose "egos were bruised" because they were not consulted or wanted a different outcome that would favour them (Holden, personal communication, 2019). According to Turton (personal communication, 2018), "anti-mining activists march to a defined drumbeat, and will always attack any proposal made by what is perceived to be the enemy (solution-seeking specialists) typically accusing them of being biased or on the payroll". This has made honest communication about complex issues very difficult and may continue to prevent progress towards a better solution. This is due to the various interests of community members who want to protect their livelihoods and businesses and focus on pursuing their interest. "In this country we are always blaming someone and we need to just fix it. Departments don't think of us the business people. It's us who need to keep the economy going and keep our people in a working position" (Pretorius, personal communication and site visit, 2018). Due to these interests, community members take matters into their own hands (as discussed in Sect. 6.3), because they believe that this is the only way to protect their livelihoods (Storey, personal interview and site visit, 2018). Nigel community member Adrian Storey claimed that his knowledge on AMD was limited. However, in his view, "I live on the river and if I see the water changing colour then I know something is going wrong, and I do not like what is happening, because this is all we have left. What we leave behind here we leave for our children and I don't believe in being a person with a short-term goal like some predominant people in our lives that look for profits". This statement depicts the environmental interest of this community member. His perceptions of the STT will be more negative because his environmental interests influence his constructions of the water quality.

With AMD treatment, no one was trying to get rich in the process (Pillay, personal interview, 2018), unlike coal mining projects that are profit oriented. The community was challenging during the EIA public participation process for the sludge disposal because there were accusations of corruption regarding the exemption of the EIA for the construction of the treatment plant and discharge of neutralised water. However, some community members had valid views that were not heard, even if they thought of valid options that the project team did not think of (Maurizi, personal interview, 2018). An example of this would be using the mine void for the sludge disposal site. How people view an EIA links to "what quality of water each person wants, who wants to drink the water and who wants it for crops" (Maurizi, personal interview, 2018). Everyone's individual interests in the catchment cannot be pleased, "what do we do, there are people, there's business and there are environmental concerns to account for. How do we take into account everyone's needs? It is not possible" (Maurizi's, personal interview, 2018). However, Humphreys suggested that the underlying problem was political will; the public understood what needed to be done in terms of the STT, but they were not heard. This is why the project took longer to be initiated and because of this when the planning started, the decisions made were attributed to a tight timeline. "People [the DWS] don't do things for the right reason, they want to do things for the most beneficial reason" (Humphreys, personal interview, 2018).

During the EIA process for the sludge disposal, Karan Beef was quite concerned about the water quality (Pillay, personal interview, 2018), but towards the end, they saw the potential use in the discharged water, due to the volumes. Therefore, having access to water would be more important in the future (van Niekerk, personal interview, 2018). Drought can influence people's views; they would rather have water than no water at all for business. This is confirmed by the fact that in December 2018 the AMD treatment plant stopped pumping water due to faults with the equipment, complaints about limited water in the Blesbokspruit surfaced (Blesbokspruit Forum 2019b). "Water is becoming scarcer, so issues over who wants it will be bigger" (Pillay, personal interview, 2018). This is similar to the view of farmers who suggested that their interest was not necessarily with having potable water, but enough water to carry the nutrients into the plant (Agricultural sector representative, personal interview, 2018), therefore suggesting that water quantity is more important than water quality, if one had to choose.

Water is scarce in South Africa and once desalination begins, people have to get used to the idea that they will be drinking treated mine water (Maurizi, personal interview, 2018). In conclusion, the implementation of the AMD STT did not take into account the community's perspectives, and therefore, the community would be less confident in the quality of potable water in the AMD LTT. The role of the DWS will be to build trust and acceptance in communities and local government (the

CoE) for the quality of water the LTT will deliver (Maurizi, personal interview, 2018). In addition, the desalinated water from the LTT might not be discharged into the Blesbokspruit for people's use but sold to big companies such as Sasol, a petroleum company. This meant that all stakeholders need to consider what is worthwhile: having more water of a lower quality or less water of a better quality (Pillay, personal interview, 2018). Activists, the tourism sector and community members might lean towards less water of a better quality, but the agricultural sector and mining industry might not feel the same (Pillay, personal interview, 2018). It remains evident that key individuals' and stakeholders' vested interests in the use of water influenced how they perceive the importance of its quality and therefore how they constructed its quality. Whose interest influences the decision-making is what links to power relations regarding AMD and its STT (discussed in Chap. 7).

6.8 Conclusion

The chapter presented the factors that influenced individuals' social constructions of water quality. Using social constructionism as the framing for this book, assisted with explaining how individuals think about relationships defined by water. These factors emerged from the interviews and are therefore not the preconceived ideas of the researcher. The commonality that existed in various individuals' understanding of the Blesbokspruit system was how the physicality of the river had changed due to human-induced events, which have contributed to the flows and volume of water. Perceptions of the AMD STT, based on the way in which it was implemented, stemmed from how the discharge of neutralised water had potential impacts on the flow and volume of water in the Blesbokspruit. Vested interests are a particularly significant factor in how water quality is socially constructed, because individuals understand the consequences of pollution from mining and therefore reach the perceptions that they do. Different interests have led to differences of opinion about the changes in the flows and volumes of water in the Blesbokspruit. This was mainly a result of how the Blesbokspruit was managed and the government's inability to enforce compliance by the mining industry and other pollution-related events that occurred over the years. Their perceptions of AMD STT developed from the community's experience with the changes in the Blesbokspruit, which was influenced by other factors such as future coal mining, reed spraying, sewage spills and water quality sampling. These factors further influenced constructions of the water quality based on the management of the Blesbokspruit. There are differences of opinion about how these changes affect individual interests, and therefore opinions and perceptions stem from the importance of these changes. This chapter demonstrated the importance of communication and information sharing. A government project such as the AMD STT in the Blesbokspruit area serves as an example of how the implementation of such a project can cause negative perceptions due to factors that influence the social constructions of the quality of water.

References

Agricultural sector representative. 2018. Personal interview, 12 April. Springs, South Africa.

Barker, A. 2018. Consultant: DRD Gold. Personal interview, 4 June. Panorama, South Africa.

Blesbokspruit Forum. 2017. Minutes Blesbokspruit Forum, 2 February. Available at: http://www.reservoir.co.za/forums/vaalbarrage/blesbok_forum/blesbok_home.htm. Accessed on: 31 January 2020.

Blesbokspruit Forum. 2018a. Minutes of Blesbokspruit Forum meeting and personal attendance, 8 February 2018.

Blesbokspruit Forum. 2018b. Minutes of Blesbokspruit Forum meeting and personal attendance, 11 May 2018.

Blesbokspruit Forum. 2018c. Minutes of Blesbokspruit forum meeting and personal attendance, 8 August 2018.

Blesbokspruit Forum. 2018d. Minutes of Blesbokspruit Forum meeting and personal attendance, 8 November 2018.

Blesbokspruit Forum. 2019a. Minutes of Blesbokspruit Forum meeting and personal attendance, 7 February 2019.

Blesbokspruit Forum. 2019b. Minutes of Blesbokspruit Forum meeting, 9 May 2019.

Blesbokspruit Forum. 2019c. Minutes of Blesbokspruit Forum meeting, 8 August 2019.

Blesbokspruit Forum. 2019d. Minutes of Blesbokspruit Forum meeting and personal attendance, 7 November 2019.

Blesbokspruit Forum Google group. 2018. Cowles Dam. E-mail communication, 14 September 2018.

Bosman, C. 2018. Consultant for Environmental health services. Personal interview, 12 April. Pretoria, South Africa.

Burr, V. 2015. *What is social constructionism?* (3rd edition). London & New York: Routledge.

DA Gauteng. 2017. DA officially objects to Palmietkuilen coal mine. Available at: https://www.dagauteng.org.za/2017/06/da-officially-objects-palmietkuilen-coal-mine. Accessed on: 10 October 2018.

De Jager, A. 2018a. Community member. Personal communication, 27 February. Springs, South Africa.

De Jager, P. 2018b. Lawyer and Springs resident. Personal interview, 25 January. Springs, South Africa.

Digby Wells Environmental. 2014. *Environmental impact assessment for the construction of the proposed sludge disposal facility and pipeline associated with the treatment of acid mine drainage from the eastern basin of the Witwatersrand, Gauteng: Full comment and response report.* Johannesburg: Digby Wells and Associates.

Federation for a Sustainable Environment (FSE). 2017. Open-cast coal mine a threat to springs wetland. *FSE website*, 3 June. Available at: https://www.fse.org.za/index.php/news/item/555-open-cast-coal-mine-a-threat-to-springs-wetland. Accessed on: 11 October 2019.

Ferreira, M. 2019. Community member. Personal communication, 16 May 2019.

Fourie, B. 2019. Aquatic Ecologist and Consultant: City of Ekurhuleni (CoE). Personal interview, 8 February. Midrand, South Africa.

GDARD Representatives (Jonhasi, C. Maluleke, R., Joshua, Q. & Mashau, B.) 2019. Gauteng Department of Agriculture and Rural Development (GDARD). Personal interview, 5 March. Marievale Bird Sanctuary, South Africa.

Govender, B. 2018. Chief directorate Mine water management unit: Department of Water and Sanitation. Personal interview, 1 March. Pretoria, South Africa.

Holden, R. 2019. Business Analyst: TCTA. Personal Communication, 12 March.

Humphreys, J. 2018. Ekurhuleni Ward Councillor: Bedfordview. Personal interview, 11 April. Nigel, South Africa.

Klip River Forum. 2018. Minutes of Klip River Forum meeting and personal attendance, 8 May 2018.

Labuschagne, W. 2018. Ekurhuleni Ward Councillor: Nigel. Personal interview, 11 April. Nigel, South Africa.

Madden, S. 2018a. Environmentalist and Springs resident. Personal interview, 8 February. Springs, South Africa.

Madden, S. 2018b. Environmentalist and Springs resident. Personal communication and Ramsar site visit, 27 February. Springs, South Africa

Madden, S. 2018c. Environmentalist and Springs resident. Personal Communication, 13 December 2018.

Maurizi, A. 2018. Manager: Water Quality Division, Ekurhuleni. Personal interview, 3 May. Springs, South Africa.

Meyer, E. 2018. Blesbokspruit Trustee Member. Personal interview, 2 February. Springs, South Africa.

Naidoo, P. 2018. Manager Platinum and Metals: Impala Platinum. Personal interview, 7 March. Springs, South Africa.

Nell, P. 2016. Researcher: Agricultural Research Council. Personal interview, 8 August. Pretoria, South Africa.

Pillay, M. 2018. Business Development Executive: Digby Wells Environmental. Personal interview, 16 January. Bryanston, South Africa.

Pretorius, B. 2018. Owner at Bush Inn and Stable Inn conference venue. Personal communication and site visit, 10 May. Springs, South Africa.

Reuter, R. 2009. Supplement conversion ratio. *Beef Magazine*. Available at: https://www.beef-magazine.com/nutrition/1104-supplement-conversion-ratio. Accessed on: 2 September 2020.

Scott, R. 1995. *Flooding of the Central and East Rand gold mines: An investigation into controls over the inflow rate, water quality and predicted impacts of flooded mines* (Water Research Report No. 486/1/95). Pretoria: Water Research Commission. Available at: http://www.wrc.org.za/wp-content/uploads/mdocs/486-1-95.pdf. Accessed on: 11 June 2019.

Springs Advertiser. 2018. Proposed Palmietkuilen Coal Mine Public Meeting. *Springs Advertiser* Facebook Group.

Stewart, S. 2019. Community member. Personal communication, 16 May 2019.

Storey, A. 2018. Nigel resident. Personal interview and site visit, 26 January. Nigel, South Africa.

Swyngedouw, E. 2009. The political economy and political ecology of the hydrosocial cycle. *Journal of Contemporary Water Research and Education*, 142:56–60.

Tlale, S. 2018. AMD project manager: TCTA. Personal interview, 9 March. Centurion, South Africa.

Tlale, S. 2019. AMD project manager: TCTA. Personal communication, 11 April.

Torregrosa, ML., de las Mercedes, D., Acosta, A. & Kloster, K. 2019. Water and society: The multidimensionality of water quality. In *Water quality in the Americas: Risks and opportunities*. Edited by The Inter-American Network of Academies of Sciences. Mexico: INAS: 20–26.

Tshikalange, S. 2020. Vaal River pollution will take at least three years to fix, says department. *Times Lives*, 23 September. Available at: https://www.timeslive.co.za/news/south-africa/2020-09-23-vaal-river-pollution-will-take-at-least-three-years-to-fix-says-department/. Accessed on: 23 September 2020.

Turton, A. 2018. Mine voids. Personal communication, 18 October 2018.

Van Blerk, M. 2019. Community member. Personal communication, 16 May 2019.

Van der Merwe, C. 2018. Chairman: Blesbokspruit Trust. Personal interview, 21 February. Springs, South Africa.

Van Niekerk, J. 2018. Consultant: Karan Beef. Personal interview, 15 February. Roodepoort, South Africa.

Van Zyl, B. 2016. General Manager: Transvaal Agricultural Union (TLU). Personal interview, 11 October. Pretoria, South Africa.

Wetzel, RG. & Likens, GE. 1991. Benthic fauna of lakes. In *Limnological Analyses*. New York, NY: Springer. Available at: https://link.springer.com/chapter/10.1007/978-1-4757-4098-1_12#citeas. Accessed on: 5 March 2020.

Wildlife Habitat Management Institute. 2005. Wading birds. Available at: https://www.nrcs.usda.gov/Internet/FSE_DOCUMENTS/nrcs143_009973.pdf. Accessed on: 16 October 2019.

Chapter 7
How Social Constructions of Water Quality Link to Power Relations Regarding AMD and Its Treatment

Abstract This chapter presents how the various factors that influenced social constructions of the water quality related to acid mine drainage (AMD) short-term treatment (STT) can be attributed to power relations, within or between society, individuals and institutions. Different types of power comprise different characteristics and are used for different purposes. These types are then either implicitly or explicitly used by various stakeholders (i.e. society, individuals and institutions) to make decisions regarding the water quality of the Blesbokspruit and AMD STT. This means that how power dynamics are linked to social constructions of the water quality depends on how the stakeholders view how management decisions are made. This is based on their interests in the water. The following elements of power were identified from the research conducted for this book and are referred to in this chapter. These include influence, authority or legal power (in a hierarchy of power), power inherent in communication and information, capacity and credibility and reputation. All these elements of power play a pivotal role in how it links to the social constructions of water quality. These elements explain how and by whom decisions are taken or actions implemented. The use of power by certain stakeholders or the absence of power of certain stakeholders indicates how those who observe these actions understand or perceive them. These perceptions are therefore a product of how various stakeholders, who are either in an observer or contributor position, understand the impact or meaning of these power elements regarding AMD and its treatment. This is what results in social constructions. The sections in this chapter rely on the logic of these elements in order to present how the individuals used references to power in their social constructions of water quality of the Blesbokspruit within the specific context to AMD and its treatment. The chapter begins by providing an overview of South African legislation and regulation and issues between the spheres of government. This includes how water use licences, environmental authorisations and mining rights have been issued in the catchment, and the power issues pertaining to these. Moreover, how power is linked to compliance monitoring and enforcement, and exemptions from environmental impact assessment processes, is explained. The issues stemming from miscommunication about the sludge disposal site and the implementation of the long-term treatment (LTT) are then discussed. The main purpose of this chapter is to indicate how water

S. Naidoo, *Social Constructions of Water Quality in South Africa*,
https://doi.org/10.1007/978-3-030-98237-9_7

management and (in)effective governance allows one to think differently about power dynamics, and how the social constructions of water quality emerge from power relations with and through water.

Keywords Social constructions · Power relations · Environmental impact assessment · Communication · Legislation

7.1 Introduction

> We should judge the previous generation. We can't judge the current generation, because they are living in the previous generation's damage. (Bosman, personal interview, 2018a)

This chapter presents how the various factors that influenced social constructions of the water quality related to acid mine drainage (AMD) short-term treatment (STT) (discussed in Chap. 6) can be attributed to power relations, within or between society, individuals and institutions. Different types of power are comprised of different characteristics and are used for different purposes. These types are then either implicitly or explicitly used by various stakeholders (i.e. society, individuals and institutions) to make decisions regarding the water quality of the Blesbokspruit and AMD STT. This means that how power dynamics are linked to social constructions of the water quality depends on how the stakeholders view how management decisions are made. This is based on their interests in the water.

The following elements of power were identified from the research conducted for this book and will be referred to in this chapter. All these elements of power play a pivotal role in how it links to the social constructions of water quality. First, *influence*, which plays a role in understanding the behaviour and decisions of others. It looks at who are the individuals that are influential and why are these individuals known to be influential and what enables them to have the ability to implement and produce results through their influence. Second, *authority or legal power* (in a hierarchy of power); this is about who has the ability to authorise an action or decision outside of legislative requirements, or what is stipulated in the law, and how this ability is attributed to the stakeholder's authoritative position or legal power. Third, power inherent in *communication and information*; this is about who has access to means of communication and access to information and their ability to use their access effectively. Fourth, *capacity,* namely those who have the human and financial resources required to implement and produce positive results will have more power. Fifth, *credibility and reputation,* indicating that stakeholders who possess these characteristics will have more power than others who do not, that may hold more senior positions. These elements explain how and by whom decisions are taken or actions implemented. The use of power by certain stakeholders or the absence of power (those not in a position of authority) of certain stakeholders indicates how those who observe these actions understand or perceive them. These perceptions are therefore a product of how various stakeholders, who are either in an observer or contributor position, understand the impact or meaning of these power elements regarding AMD and its treatment. This is what results in social constructions.

The sections in this chapter rely on the logic of these elements in order to present how the individuals used references to power in their social constructions of water quality of the Blesbokspruit within the specific context to AMD and its treatment. The chapter begins by providing an overview of South African legislation and regulation. The purpose is to understand the law and legal requirements pertaining to the use and protection of water resources and the environment in order to present the difference in the legal requirements when a project is implemented under emergency measures. The reason for presenting South African legislation and regulation in this chapter is because it was suggested by some stakeholders' that the government did not abide by the law when implementing the STT of AMD. Then this chapter explains the issues between the spheres of government identified in the findings and are discussed in four subsections to provide stakeholders' understanding of the roles and responsibilities of government departments in managing the Blesbokspruit. This includes how water use licences, environmental authorisations and mining rights have been issued in the catchment and the power issues pertaining to these. Moreover, how power is linked to compliance monitoring and enforcement and exemptions from environmental impact assessment processes is explained. The issues stemming from miscommunication about the sludge disposal site and the implementation of the long-term treatment (LTT) are then discussed. The main purpose of this chapter is to indicate how water management and (in)effective governance allow one to think differently about power dynamics and how the social constructions of water quality emerge from power relations with and through water.

7.2 Legislation, Regulation and Governance

> Good governance means having processes, and setting examples, so things are done properly. (Mashau, GDARD representative, personal interview, 2019)

The purpose of this section is to present the required role of the South African government to protect the country's water resources, prevent pollution of the environment and ensure remediation takes place where environmental degradation has occurred. This section examines the institutional roles of government stipulated in four acts presented below that were relevant for the AMD STT, although a different process occurred once implemented. The relevance is because key individuals from the community, local government, private sector, tourism industry, researchers and activists, according to their knowledge of the law, were of the opinion that the government had not abided by legislation and regulations when implementing AMD STT, which contributed to the way they viewed the STT process (Barker, personal interview, 2018; Bosman, personal interview, 2018a; De Jager, personal interview, 2018; Liefferink, personal interview, 2016; Madden, personal interview, 2018; Maurizi, personal interview, 2018). An overview of the Acts, policies and official government documents is provided as an indication of legal power that was relevant for this book and more specifically for discussions to follow in this chapter.

The following four Acts are important for this purpose:

1. The National Water Act (NWA) (Act 36 of 1998) aimed, after the end of apartheid, to provide fundamental reform of the law relating to water resources. It provides for matters connected therewith: (1) it acknowledges that the national government has overall responsibility for, and authority over, the nation's water resources and their use, including the equitable allocation of water for beneficial use; (2) it recognises that the ultimate aim of water resource management is to achieve the sustainable use of water for the benefits of all users; and (3) it recognises that the protection of the quality of water resources is necessary to ensure sustainability of the nation's water resources in the interests of all water users (RSA 1998a, p. 1). Due to the AMD STT being implemented as emergency, the interests of all users were not taken into account, and legal power was used to implement the treatment without an environmental impact assessment (EIA) being conducted for the sludge disposal site, and an EIA and a water-use licence for the discharge of neutralised water does not exist.
2. The National Environmental Management Act (NEMA) (Act 107 of 1998) serves as the general framework within which environmental management and implementation plans must be formulated and adhered to (RSA 1998b). Chapter Three of NEMA states that there must be procedures for co-operative governance, environmental implementation plans and management plans. The purpose of environmental implementation and management plans, according to section 12(a), is to coordinate and harmonise the environmental policies, plans, programmes and decisions of the various national departments that exercise functions that may affect the environment or are entrusted with powers and duties aimed at the achievement, promotion and protection of a sustainable environment and of provincial and local spheres of government. This policy has relevance for how water use licences, environmental authorisations and mining rights are legally issued by the responsible departments, instead of being conducted jointly to assess the impact on the Blesbokspruit. There is a lack of communication between the various government departments (discussed in this chapter) and the implementation of the AMD STT was not communicated openly with relevant government departments that are involved in managing the Blesbokspruit catchment.
3. The Minerals and Petroleum Resources Development Act (MPRDA) No 28 of 2002 (RSA 2002) aims to give effect to the principle of the State's custodianship of the nation's minerals and petroleum resources and to promote economic growth and mineral and petroleum resources development in the Republic. According to section 2(h), which is of relevance to this book, the MPRDA should give effect to section 24 of the Constitution by ensuring that the nation's minerals and petroleum resources are developed in an orderly and ecologically sustainable manner while promoting justifiable social and economic development. Section 3 states that the Minister has the ability to refuse, grant, control,

administer and manage permission of mining rights. While extensive rehabilitation in the form of the STT of AMD is taking place, mining rights are being issued in an already sensitive environment and can further impact on the water quality of the Blesbokspruit.

4. The National Environmental Management Waste Act 59 of 2008 (RSA 2008), Chapter 3 section 7(1) (a–b) states that waste must not be mixed or treated in a manner where it could result in treatment that is not controlled and not permanent. According to Chapter 3 section 7(2) (a–b), waste may be pretreated to enable potential for reuse, recycling, recovery of treatment or reduce the risk associated with the management of the water. The relevance of this act is that the AMD project team suggested that the STT will discharge an improved quality of water into the Blesbokspruit compared to the quality of raw AMD flowing into the river and through legal power implemented the STT under emergency measures.

The Mine Water Management Policy (draft policy of 2017) deals specifically with AMD and therefore has direct relevance to this book. In 2016 the Director General gave an approval for a Mine Water Management Unit to be formed within the Department of Water and Sanitation (DWS), which is led by Marius Keet (Chief Director: DWS). Bashan Govender (Assistant Director) directly supports him in this undertaking. The direct function of the unit is to provide high-level oversight and integrate efforts across the department's provincial and national offices to prioritise issues and decide what interventions are required (Govender, personal interview, 2018). Further, through the integration of a departmental approach to mine water management, the reason for this policy is to rectify where legislative gaps exist. The policy aims to provide relevant and appropriate legislative remedy in order to strengthen a proactive mine water management approach, for prospective, operational and historical mining activities (DWS 2017, p. 2). The aim of the policy is to improve water quality in South Africa (Liefferink, personal interview, 2016). The draft policy has taken into account the above acts. According to NWA (RSA 1998a), the DWS, who is the custodian of South Africa's water, has to prevent pollution through managing the land and use of the water resource. Further, as stipulated in NEMA, the DWS should ensure "duty of care and remediation of environmental damage". Thus, if a person fails to take the measures required of him or her under the Act, the action should be taken by the custodian to hold him or her accountable and prevent further ecological degradation. The MPRDA assists Ministers with recovering costs in the event of urgent remedial measures taken to protect the environment (DWS 2017, pp. 3–4). The Minister of the DWS approved the Gazette (vol. 625 no. 40987) for the Mine Water Management Policy in July 2017. A draft policy that has not been finalised is available.

According to Liefferink[1] (personal interview, 2016), "the Mine Water Management Unit is admirable". She indicated that the Mine Water Management policy aimed to respond to the practice of mining companies declaring liquidation

[1] M. Liefferink is the CEO of the Federation for a Sustainable Environment, an NGO.

to avoid environmental liabilities, and mines then becoming abandoned, with companies not applying for closure certificates. The liabilities of pumping and treating the water are then carried by neighbouring mines or, alternatively, by the municipalities and if no treatment takes place, then communities carry the cost. Therefore, the closure process needs to be amended to cater more for the environment and communities. In November 2015 the Department of Environmental Affairs (DEA) incorporated into existing regulations financial provisions which not only cover future mining companies but also those currently operating. The aim is to ensure financial contributions from mining companies for as long as they are in operation to account for latent impacts, including the pumping and treatment of acid mine water (Liefferink, personal interview, 2016).

Another relevant government document for this book is the National Water and Sanitation Master Plan (DWS 2018). The plan aligns municipal infrastructure to ensure water security and equitable access to water and sanitation (DWS 2019). The plan acknowledges that infrastructure is weak and additional support to local government is required. This is relevant for the discussion about spheres of government (discussed below): the national government's planning and implementation of AMD STT and local government's responsibility for monitoring the quality of the water. Hoekstra et al. (2018, p. 7) strongly emphasises the significance of good governance in achieving water security and having an adequate water system status, which requires integrated analysis, planning and logical policymaking across the government departments concerned. Natural scientists are responsible for designing effective solutions but they often underestimate the process needed to implement those solutions. While political scientists have the responsibility of worrying about policy processes, stakeholders and power relations and thus looking for ways to achieve good governance, they often underestimate the quality and effectiveness of the policy outcome (Hoekstra et al. 2018, p. 7).

Institutional problems in government with policy implementation have already been well recorded. This applies to the Master Plan and related policies, which elude to the effective functioning of the power hierarchy of government institutions. A reason for policy outcomes being ineffective is because the person that holds the position of Minister of Water does not always understand water[2] (Barker, personal interview, 2018; Bosman, personal interview, 2018a). This means that there are better ways that the resource can be managed. The problem with South Africa's administrative model and structure is that a director of a national government department is expected to cover the whole portfolio. Bosman[3] (personal interview, 2018a) suggested that the problem is that there is not a skilled-enough person high enough to

[2] Since 2017 the Minister responsible for water affairs changed from Nomvula Mokonyane to Gugile Nkwinti to Lindiwe Sisulu in 2019 and since 2021, Mr. Senzo Mchunu. Some research participants shared a view that if one Minister failed, he or she is replaced with someone else who is not an expert. Stakeholders were of the view that experts should be involved in such serious issues such as AMD.

[3] C. Bosman conducts training on water use licence applications and all relevant policies relating to water use.

deal with these issues on a strategic level, and some are promoted without obtaining the required years of experience or because they are good with technical issues. In addition, there were concerns around whether the DWS was undergoing financial difficulties due to media reports suggesting contractors hired by the DWS had not been paid (*Business Tech* 2017). Statistics from an audit report were publicised, illustrating significant wasteful expenditure by the DWS. In a period of 8 years, the DWS Chief Financial Officer changed several times (Blesbokspruit Forum 2018a).

Owing to these financial difficulties being publicised, concerns about funding for the LTT of AMD surfaced and whether the STT would continue despite the fact that the public were informed that it was a short-term measure. Such measures and strategies are often discussed but the outcomes are not visible. There should be stronger legal support and dependence on the private sector (Barker, personal interview, 2018). Strategies should be reviewed by people who are experts in the field; they should form part of the "task team" (Barker, personal interview, 2018). Key individuals shared common ground on the fact that the Minister of Water is not always water expert and this is part of the problem with policy. Including experts in decision-making means adding value towards addressing the water crisis (Barker, personal interview, 2018). Professor Craig Sheridan (cited in Mouton 2018) substantiated this view: "[I]n order to find sustainable solutions to South Africa's water problems, all role-players in the water sector should start to work together". There needs to be a "democracy of discipline values", where policy, engineering, science, social science, law and politics are all treated equally. Water governance will continue to fail and prolong the negative water quality in South Africa if there is no agreement on how government and businesses will respond positively and together implement smart, anti-pollution strategies stemming from AMD (Mpofu et al. 2018, p. 80). Thus, sustainable solutions can only be attained if water governance is approached in a holistic manner.

Various stakeholders were of the opinion that issues between spheres of government led to inadequate management of water resources. It was suggested that this was a result of the government not abiding by the law, which contributed to the deterioration of the water quality of the Blesbokspruit. Uncertainties about whether compliance with legislative requirements influenced constructions of water quality and their links to the power relations between spheres of government are discussed next.

7.3 Issues Between and Within Spheres of Government

> The levels of government do not know what they are doing. (De Jager, personal interview, 2018)

The Constitution refers to co-operative government where co-operation between the three spheres of government must take place in mutual trust and good faith, through supporting, assisting and informing one another, and consulting with one another on matters of common interest, by coordinating their actions with one another and following agreed procedures (RSA 1996, p. 21). However, inconsistencies between the spheres of government were commonly raised by key individuals. The main concern was with a lack of communication – a main factor influencing social constructions of water quality – between and within the spheres of government and then between government and the public.

In this book, the specific concerns raised by stakeholders were at national level between the DWS, the DEA and the Department of Mineral Resources (DMR), the three departments that have a role with AMD treatment. Oelofse (2010) illustrated the constitutional requirements between these three national departments. The DEA is responsible for wastewater management policy and therefore for providing leadership and guidance regarding wastewater to enable other national departments, provincial environmental departments and municipalities to meet their executive obligations in respect of waste management. The DWS is responsible for the protection of water resources and is therefore concerned with possible impacts from waste management practices, because the disposal of waste is listed as a water use in section 21(g) of the NWA. Therefore, the DWS has a mandate to set regulations on waste disposal and any treatment to protect water resources. Once the water use licence is issued, units within the DWS such as Water Regulations Services and Compliance Monitoring and Enforcement (CME) have to ensure the limits of the licence issued are adhered to and, if not, their responsibility is to enforce compliance (DWS 2017, p. 3). Licences are issued for a specific period, with rules and regulations regarding the quantity of water discharged and the rehabilitation processes. The DMR has waste-related responsibilities, including mining waste (Oelofse 2010).

As regards AMD STT, the DWS is responsible for the planning and implementation of the STT, the DEA has to approve and issue the environmental authorisation required for the DWS to conduct the treatment, and the DMR has to consider mining applications while the STT is taking place. This indicates the significance of co-operative governance where co-operation between departments was key for the treatment of AMD. In the Blesbokspruit, national and provincial (GDARD) government departments delegate responsibilities to local government (the City of Ekurhuleni (CoE)), but policy formulation, legislation, regulation and distribution of funds remain the authority of national and provincial government. The CoE is responsible for providing waste management services and management of waste disposal facilities. According to Oelofse (2010), if the relevant sphere of government does not have the resources or capability, then the next sphere must implement the function.

Further, the various national and provincial government departments and the local government have to work together closely. For example, section 2(2) of the NEMA states that environmental management must be integrated; it must take into account the effects of decisions on all aspects of the environment and all people in the environment by pursuing the selection of the most feasible environmental option (RSA 1998b). Ensuring integrated environmental management is how co-operative governance will be achieved and vice versa. Therefore, relevant government departments have to monitor and manage the catchment, keeping in mind all the activities that take place within the catchment, the users of the water, including the extensive and costly rehabilitation in the eastern basin, such as AMD treatment. If new mining applications are received, the DMR has an equal role to ensure the aims of co-operative government are achieved.

If there is a strong consideration for co-operative government as stipulated in the Constitution, and environmental management as stipulated in the NEMA, then the DWS's role as the custodian of water resources means that they should share concerns with the public, take their suggestions into account and discuss potential solutions. The DWS, the DEA and the DMR together with Gauteng Department of Agriculture and Rural Development (GDARD) and the CoE are meant to ensure good governance, by following good processes, policies and a legislative framework, in order to do proper work (Mashau, GDARD representative, personal interview, 2019). Govender (personal interview, 2018) acknowledged that communication between tiers of government could be weak. Co-operation between government departments should not be seen as a conflict of interest but rather to enhance information sharing and communication to promote effective governance.

Stakeholders' awareness of inconsistencies between and within spheres of government was evident when the Blesbokspruit was placed on the Montreux Record. Springs resident, De Jager (personal interview, 2018), illustrated "the levels of government do not know what they are doing". Community members are of the opinion that the Blesbokspruit is a threatened site because government officials involved did not take responsibility for it (Van der Merwe, personal interview, 2018) and did not follow the specific regulations as stipulated in the NWA (Madden, personal interview, 2018). How water is understood by people, governments and businesses played a significant role in how the Blesbokspruit is managed. This indicated how water is manipulated, used and concentrated socially by stakeholders and how the struggles for water access and control and exclusion of some stakeholders are expressed in institutions. The constructions of water quality discussed in Chap. 6 hinted at the existence of power relations through influences such as management and practices.

The subsections that follow indicate how the constructions were influenced by power relations, for instance, issues of power relationships exist between leading government departments and supporting departments or between government departments at the same level but use different power elements to make decisions. National government departments are regarded as more "senior" with more capacity and authority than provincial or local government departments. However, there are cases where a provincial department is more effective than a national department and therefore will effectively have more power in terms of influence and information (and a better reputation) in the view (or perception) of other stakeholders.

7.3.1 Roles and Responsibility of Government Departments Within the Blesbokspruit

> The law does not concern itself with trivial matters, if there is no significant impact then don't bother. (Bosman, personal interview, 2018a)

Debates over the responsibilities of government departments (and therefore their power relations) within the Blesbokspruit constantly surfaced and played a role in how social constructions of the water quality were formed. When government representatives concerned report the actions taken for their responsibilities during meetings of the Blesbokspruit Forum and the Gauteng Wetland Forum, there was often confusion over their roles and responsibilities (Blesbokspruit Forum 2018a). This resulted in some departments carrying the burden of other departments for not fulfilling their obligations. This could be attributed to the existence of power relations between the different levels of authority.

Govender (personal interview, 2018) asserted that as a government employee, irrespective of whether one is employed in a national, provincial or local government function, it is the official's responsibility to know what his or her function and mandate is and to act according to those responsibilities. "I don't believe that someone working in the municipality should be able to say, for example, that a wastewater treatment plant is not working". Local government has a duty to ensure the plant is working, serviced, financed and budgeted for, and if there is a breakdown, a mitigation plan should be in place. Where uncertainty exists, the matter should be escalated for guidance and clarity (Govender, personal interview, 2018). However, after escalation, there were still challenges that persisted at the local government level. Ward Councillors Jill Humphreys (personal interview, 2018) and Wollaston Labuschagne (personal interview, 2018) expressed the difficulty that local government had in communicating with provincial and national departments, which resulted in councillors relying on their counterparts in the provincial departments and Parliament.

Further to this, they indicated that capacity challenges existed. Too few people may be working on a portfolio committee and only certain issues can be addressed. Ward councillors mostly rely on municipal departments to relay information, challenges and concerns to the provincial and national departments, but this does not happen (Labuschagne, personal interview, 2018). Humphreys (personal interview, 2018) suggested that these inconsistencies boiled down to individual behaviour of officials from the municipality, "are they going to be seen in the right light if they report this [to provincial or national], who above me will judge me if I report this". This was justified through instances where emails were completely ignored, despite stating issues factually (Humphreys, personal interview, 2018). If local government reported a problem and no action is taken, it can be escalated to the Department of Co-operative Governance and Traditional Affairs (COGTA), the South African Local Government Association or provincial offices[4] of the DWS (Govender, personal interview, 2018). These departments are allocated funding from national government to assist municipalities, for instance, with wastewater treatment works. Therefore, Govender advised that there were processes to allow for problems to be escalated upwards until national government was informed, but the onus was at

[4] Every provincial office has a section providing water support, largely dealing with potable water supply and sanitation issues.

local government level. The national government's role is not to directly provide drinking water and sanitation but to provide high-level oversight, establish mechanisms and create the framework to allow local government bodies to achieve a service delivery function that is not vested in the national office (Govender, personal interview, 2018). This indicated a difference in opinion over how accessible spheres of government are to one another. Key individuals from national government feel that other spheres of government should know what their roles are and also have the option to escalate the matter if unaddressed. However, individuals from local government are "fearful" of reporting matters for being viewed in the incorrect light. This indicated that water management is rooted in deep political processes and such power dynamics stem from legal power in terms of capacity, communication and authority.

In addition to how accessible departments are to one another, representatives from the various departments are not always present at the catchment forums to report back on their portfolios, creating lack of awareness about what each department is doing (Fourie, personal interview, 2019). This led to the development of an "action column" added to the Blesbokspruit Forum agenda to ensure relevant departments report back on what they were responsible for (Blesbokspruit Forum 2018b). The aim was to enhance the transfer of communication and information. However, this still did not encourage representation of some government representatives. Forum representatives, such as those from Rand Water and the private sector, were vocal about the lack of commitment from the various government departments (Blesbokspruit Forum 2018c). For instance, a representative of the DMR was seldom present at the forums and it became a topic of discussion (Blesbokspruit Forum 2018c, d, 2019a, b, c, d, 2020a, b). The lack of commitment from the DMR to be present and report back at the forums created difficulties for other government departments to fulfil their roles and understand what the cumulative impact on the Blesbokspruit was. This also put into question the DMR's credibility to fulfil its management role of the Blesbokspruit.

Managing the catchment is challenging when organisations do not provide information during the forums regarding their direct discharges or when new mining rights are issued (Maurizi in Blesbokspruit Forum 2018d). During the Blesbokspruit Forum (Rand Water representative in Blesbokspruit Forum 2018d), it was suggested that the CoE take an altruistic role in managing the Blesbokspruit catchment. However, the CoE requires all water quality results to be submitted accordingly in order to fulfil its role (Maurizi, personal interview, 2018). With urbanisation there is a constant flow of water and people have mixed views about discharging water back into the river, "but what then, where must we put it. If the people are removed then there won't be water to discharge" (Maurizi, personal interview, 2018).

The differences in the use of the water make it difficult to manage the water more effectively. For instance, GDARD is involved in managing the Marievale area, which is in their jurisdiction, but only spray their demarcated areas for reeds. Mines point fingers and blame one another for the poor water quality discharges. The aim should be to create solutions not assigning blame (Fourie, personal interview, 2019). This showed the power relations between departments, due to the DMR's lack of presence at the forums meant that matters regarding mining rights cannot be explained. The DMR exercises their legal power by issuing a mining right but does not communicate their decisions to other departments who are involved in managing the water quality of the Blesbokspruit.

Diversity in activities implies that coherence in decision-making requires visibility between local, provincial and national government (Blesbokspruit Forum 2018c; Bosman, personal interview, 2018a; Gauteng Wetland Forum 2018a). For instance, new initiatives are presented at the forums from various government departments to keep one another informed and to provide a platform to identify issues encountered with their implementation. GDARD (Blesbokspruit Forum 2018c) is working on a river health programme. The purpose is to assist policymakers to make more informed decisions with regard to managing river health (Blesbokspruit Forum 2018c). The DWS initiated a Wetland Monitoring Programme, which is meant to identify owners of the land around the wetland, the level of protection that the wetland requires and the indicators relevant when applying for a water use licence and environmental authorisation (Gauteng Wetland Forum 2018b). However, lack of staff capacity within the DWS meant that implementation of the programme will take longer (Gauteng Wetland Forum 2018b). If such initiatives worked, there would be a holistic management approach of the Blesbokspruit catchment (Fourie, personal interview, 2019). Considering the diversity of activities requires the roles and responsibilities of government departments to be aligned to address issues in the catchment from a "cumulative perspective" (Barker, personal interview, 2018). The vision of the Mine Water Management Draft Policy is to clarify the delegation of powers between different spheres of government (DWS 2017, p. 5). The need for clarity of roles was already raised by stakeholders in provincial (GDARD) and local government (CoE and ward councillors) during the EIA process for the sludge disposal site (Digby Wells, Environmental 2014).

Clarity of roles between government departments is required for improvement in the management for the Blesbokspruit. Bosman (personal interview, 2018a) acknowledged that pollution could not be fixed overnight, and van Niekerk[5] (personal interview, 2018) suggested that it was due to South African legislation being too demanding; "our water rules and regulations are very sophisticated and too cumbersome, that everything cannot be addressed". However, Mashau[6] from the GDARD said it was a matter of good governance, which means having processes and setting examples so that processes run accordingly. Each activity that takes place in the Blesbokspruit has an application process and inspections need to be conducted (Mashau, GDARD representative, personal interview, 2019). Mashau suggested that it would be beneficial for inspections, monitoring and evaluation to take place to identify whether activities are conducted in a lawful manner. Therefore,

[5] J. Van Niekerk is a consultant at Karan Beef Feedlot and represents the Feedlot at the Blesbokspruit Forum.

[6] B. Mashau is the Project Administrator for the Blesbokspruit Ramsar site at the GDARD.

through improved governance, the management of the Blesbokspruit can be strengthened.

Each government department wants to be involved in decision-making to ensure that specific interests are considered in the way decisions are implemented and managed, as in the case of AMD STT. Existing issues with water governance led to the STT process being questioned. Which resulted in the public viewing it as a political issue between government departments. Many suggestions were made during the EIA process about how management should take place. For instance, it was suggested that COGTA should lead the AMD treatment process and local authorities should form part of the decision-making body on AMD matters in the province (Digby Wells Environmental 2014). However information on the STT was not shared openly, and Barker[7] (in Klip River Forum 2018) suggested than an AMD Awareness Programme be developed to improve information sharing to align the roles of the DWS and that of the CoE who has to limit the water pollution. This indicated the need for transparency in information and communication. However, this implied that there was implicit power dynamics regarding the STT of AMD, linked to lack of communication and information sharing.

In Bosman's (personal interview, 2018a) opinion, different bodies involved in the water sector do not communicate with one another and do not share information openly because the politics and money involved in projects are bigger than what they appear at face value. The DWS is addressing the environmental problem but there is immense criticism from individuals in the community, activists and local government due to the way the problems are managed and a lack of communication between spheres of government and then with the public (Blesbokspruit Forum 2018a; Bosman, personal interview, 2018a; Labuschagne, personal interview, 2018; Madden, personal interview, 2018). This is further justified by the fact that GDARD completed the application for the Blesbokspruit to be removed from the Montreux Record early in 2019 (Blesbokspruit Forum 2019b). The DEA is responsible for submitting the application to the Ramsar Convention. At the end of 2019 (Blesbokspruit Forum 2019d), the DEA had still not submitted due to unknown administrative issues. Stakeholders (Blesbokspruit Forum 2019d) voiced their frustrations and complaints to the GDARD, even though GDARD had fulfilled their role. This is an example of power relations that lead to conflict stemming from differences in the authoritative power between spheres of government that can arise and impact on the efforts of one department due to the delays in another (GDARD representatives, personal interview, 2019). The DEA has the authoritative power to submit the documents required to remove the Blesbokspruit from the Montreux Record. However, the GDARD has power in terms of their credibility to complete the application in a timely manner. Questions raised are directed at the GDARD but they do not have the authoritative power to see the process through. This could have resulted from challenges associated with unclarified roles and responsibilities, which is evident of poor governance.

[7] A. Barker is a consultant for Ergo Mining and represents this mine at the Blesbokspruit Forum.

The discussion in this section demonstrated how the various issues regarding roles and responsibilities among government departments result in constructions of the water quality which are linked to power relations.

7.3.2 Issuing of Water Use Licences, Environmental Authorisations and Mining Rights

> If it's not a law in the act, then it doesn't stand for in the water use licence application. The NWA clearly stipulates what a water use is. (Bosman 2018b)

> Does the mine put their green hat on when applying for a water use licence? (Bosman 2018b)

This section explains the issues and differences in opinions that exist between spheres of government from three perspectives: (1) between national government and local government, when national government makes decisions to issue water use licences without informing local government; (2) within national government when issuing water use licences, environmental authorisations and mining rights; and (3) between national government and the public, in terms of how AMD treatment was granted an emergency status. The issues and differences that exist between spheres of government influence the social constructions of the water quality because of how they are linked to power relations.

The DWS is responsible for issuing a water use licence but monitoring of pollution incidents is under the CoE's jurisdiction. The CoE therefore has to deal with the consequences of the DWS's decision to issue the licence; Maurizi (personal interview, 2018) finds this process complicated in her role at the CoE. In the past meetings were held between the DWS and the CoE to inform all relevant parties of all water use licenses issued, but this no longer happens. The CoE has to monitor the water use and discharges, and if any discrepancies or impacts are visible from the water samples, they are kept as evidence because it is the CoE's responsibility to monitor the water (Maurizi, personal interview, 2018). Further, the DMR grants mining rights, and because some of these applications are granted and not communicated at the forum, monitoring and measuring of the overall impacts cannot take place. Therefore, no knowledge exists of whether a particular user is exceeding the level of their water use (Blesbokspruit Forum 2019c). The problem is not only when a specific mine is not adhering to its water use licence but also when the DMR and the DWS do not follow up on whether those conditions were met, which results in public frustration developing (Maurizi, personal interview, 2018). The CoE has monthly meetings regarding EIAs and a representative from the DWS is present, but there are different levels of communication and very little information is provided (Maurizi, personal interview, 2018). The Trans-Caledon Tunnel Authority (TCTA) is monitoring all water users in the area due to AMD STT discharge; therefore, information on both the upstream and downstream discharges are available to the

CoE through these means. Everyone is trying to do his or her job as best as possible, but improved communication would assist (Maurizi, personal interview, 2018).

Lack of communication leads to contradictions as to why the DWS issues water use licences in an area that is undergoing excessive rehabilitation and discharge into the water course. The problem was that the DMR issues authorisation for a mining right without the mine having a water use licence, which puts pressure on the DWS to issue the water use licence (Gebhardt in Blesbokspruit Forum 2018d; Tlale, personal interview, 2018a). This hints at a problem with the order of how licences and authorisations are issued, stemming from the DMR, the DEA and the DWS working in isolation. The public are of the opinion that this is due to organisational interests and institutional affiliation (Blesbokspruit Forum 2018d). The three licence applications should be done together instead of separately (Barker, personal interview, 2018). This problem is illustrated in the case of the Palmietkuilen Mining operation (open-cast coal mine). According to the FSE (2019), the Mine Water Management Policy refers to open-cast coal mining as having acid-generating potential. The mine received a water use licence but the public were not informed; this is an infringement of their rights and they have the right to appeal the licence (FSE 2019). This contradicts the DWS opposing the DMR issuing mining rights in an area undergoing extensive rehabilitation because the DWS still issued the water use licence (De Jager, personal interview, 2018). An influx of applications should not be a concern but rather a thorough assessment, determining the acceptance or rejection of the application (Blesbokspruit Forum 2018d), because development and business cannot stop completely for the environment or vice versa (GDARD, in Blesbokspruit Forum 2018d).

There are different elements of powers, such as authority/legal power and credibility and reputational power, which are aligned to the roles and responsibilities of the different government departments. These different roles and responsibilities create power relations. For instance, the GDARD is strictly linked to protecting the environment, and this can prevent the DMR from achieving its objectives of sustaining mineral resources as required in the MPRDA. From a public point of view, it will seem as if there is "internal conflict" (or a power struggle) between government departments, because one department has mining interests and the other environmental interests (Jonhasi, GDARD representative, personal interview, 2019). But it is their governmental and institutional responsibility to focus on those interests. However, the aim is to achieve a balance, and "green organisations" (environmental activists) will require a zero level of pollution, but government (the DMR, the DWS and the DEA) will allow certain levels of pollution (Jonhasi, GDARD representative, personal interview, 2019). To an outsider, it would seem as if government departments responsible are allowing pollution to occur, but this is not the case. This means that a "trade-off" has to be achieved in terms of the socio-economic situation in the country said Jonhasi. At this stage, it is important to determine what is or is not considered pollution. According to Bosman (personal interview, 2018a), the Constitution refers to pollution as causing harm to people, meaning the water is unsustainable. As regards the AMD STT, Bosman strongly believed that if the Minister signed off the current disposal of a sludge site, it does not mean that it is "legal". The DWS should implement all reasonable measures such as obtaining a water use licence to identify the impact on the water course and how pollution will be prevented. The research findings indicate that this is why it is viewed as "conflict" that exists between spheres of government responsible for different sectors (Jonhasi, GDARD representative, personal interview, 2019). This perceived "conflict" can lead to an exertion of power to fulfil different stakeholder interests.

Interests are fulfilled and decisions are made, for instance, because unemployment levels are high. In such an instance, the environment will be "traded" for employment by some, but for others, such as the community, the environment should come first, meaning the reeds should be maintained and the bird life should thrive (Mashau, GDARD representative, personal interview, 2019). Therefore, the view that government departments are allowing pollution arises because the community and government interests differ. The DMR and the DEA will be pressurised from communities due to socio-economic factors, and the public is not always aware that government officials have agreed on the level of environmental impact, or have agreed on a "trade-off", because of the different interests that exist (Jonhasi, GDARD representative, personal interview, 2019). Some departments receive different instructions from their respective Ministers, and this results in no action being taken, which leads to conflict due to disagreements on a way forward. If each department had to put its interests at the forefront without communicating with one another, there would be no consistency at all (GDARD representatives, personal interview, 2019). This means that these competing interests influence constructions of water quality, and this discussion demonstrated how these interests are linked to power relations based on fulfilling institutional interests.

An example of such inconsistency is that of Canyon Coal's Ukufisa Colliery, which received all necessary authorisations in 2017. In March 2019, the Federation for a Sustainable Environment (FSE) reported – through the Blesbokspruit Forum – evidence of the DMR's lack of enforcement of non-compliance by this colliery. "It's been going on for a long time and it is not a legal operation; it was legalised in an illegal way" (Van der Merwe, personal interview, 2018), because the mine is comprised of former directors from the DMR (De Jager, personal interview, 2018). The implication is that someone is making it easy for mines to get their way (Humphreys, personal interview, 2018). There are many "illegalities" in current mining activities (Labuschagne, personal interview, 2018). However, in the case of Canyon Coal's Palmietkuilen Coal Mine, the owner publicly stated that allowing the mine to operate would assist Eskom with power supply and that the operations would be conducted according to the terms stipulated in the water use licence and environmental authorisation (Creamer 2019). However, through an ongoing public appeal (Blesbokspruit Forum 2018b; Labuschagne, personal interview, 2018; Tlale, personal interview, 2018a) from the Blesbokspruit Trust, community members, the FSE and the Wildlife and Environment Society, the environmental authorisation was overturned by the Minister of Environmental Affairs, Barbara Creecy, based on the motivated impact that it would have on agricultural land (Tempelhoff 2019). This is an example of how coal mining that influences individuals' social constructions of water quality is linked to power relations. This is an example of power through influence by those stakeholders who are not in an authoritative position.

Influential stakeholders had the ability to convince the Minister of Environmental Affairs, who was in a position of power based on authority, to change the decision.

Cases such as the Palmietkuilen Mine where authorisations were revoked after being issued due to power being exercised by influential stakeholders can be addressed through improved communication at national government level. Agreements should be reached before issuing any licence or authorisation. A suggested strategy was for the DMR to present the physical outlook and map of abandoned tailings facilities/mine dumps and current and new mining applications to forum members (Blesbokspruit Forum 2018c, d, 2019a). This would enable the DWS to make the water use licence process more efficient and at the same time not compromise the quality of the process (Blesbokspruit Forum 2018d). However, because a DMR representative was not present at the forums on a regular basis, inconsistencies persisted (Blesbokspruit Forum 2018c, d, 2019a, b, c, d). The DWS, in turn, created an electronic system for licence applications to be distributed among stakeholders as requested, so that all water use licences are transparent (Blesbokspruit Forum 2018a, b), and presented and discussed at the forum (Barker in Blesbokspruit Forum 2018c). However, there was still room for improvement from the DWS. Some water use licences were not available and therefore the use of the licence was not publicly known, or whether conditions were adhered to, which stakeholders found to be unacceptable (Blesbokspruit Forum 2018c). Further, stakeholder concerns stemmed from these inconsistencies. A member of the community and Rand Water felt that the government only looked at the environmental impact of individual applications, instead of the increasing impact from an increased use of water. A Strategic Environmental Assessment, the purpose of which is to assess all environmental authorisations conducted to present a "cumulative impact", could cover this (Tlale, personal communication and site visit, 2018b), creating an easier flow of information (in Blesbokspruit Forum 2019b). These inconsistencies raised the fact that environmental consultants may not have access to necessary information and therefore provide inaccurate information to the decision makers when conducting an EIA (Blesbokspruit Forum 2019b). This led to the suggestion that the public be part of the water use licence application and environmental authorisation process, as in the case of the AMD STT process, which the current structures do not allow for (Blesbokspruit Forum 2019b).

Key individuals found it to be unacceptable that a water use licence for the AMD STT did not exist (Bosman, personal interview, 2018a; Pillay, personal interview, 2018). According to their understanding, it is a legal requirement. "The AMD STT theoretically needed a water use licence for the treatment and abstraction of water but because the DWS was the applicant, they claimed they could not issue themselves a licence to discharge the neutralised water or disposal of sludge into the mine void" (Pillay, personal interview, 2018).

The disposal of sludge forms part of the listed water uses under section 21 of the NWA. Promotion of Justice Act (PAJA) (Act 3 of 2000) states that "the minister cannot delegate authority to issue a water use licence to the department undertaking the activity, only the minister can issue it" and the licence must be made publicly available if requested (Bosman, personal interview, 2018a, b; DWS representative in the Blesbokspruit Forum 2019a). However, Govender (personal interview, 2018) suggested that a water use licence was not required as the TCTA had implemented the AMD project under section 110 of the NWA, as a government waterworks. Further, the Legal Services department at the DWS confirmed that the DWS cannot issue a water use licence for a government waterworks implemented by its entities (the TCTA) (Blesbokspruit Forum 2017a). Further, according to legislation (Govender, personal interview, 2018), a water use licence can only be issued once an environmental authorisation has been granted. In the case of the STT this was not needed because the project was granted an emergency status, which removes the requirement for an environmental authorisation and therefore the need for a water licence. Pillay (personal interview, 2018) felt that the purpose for a water use licence cannot be satisfied through an EIA. This discussion indicated the confusion that existed over the water use licence for the STT and stemmed from a lack of communication and different interpretations of procedures and laws. The public feeling was that the national government does not need to abide by the same processes as other users of the water. The way the STT was implemented influenced constructions of water quality. What some individuals understand about legislation, in their opinion, was not practised during the implementation of the STT. The research identified that for these individuals it was due to the DWS's use of its authoritative power to exempt itself from requiring a water use licence for its activity. This is therefore an example of how the use of power, in a legal manner, negatively influenced social constructions of the water quality. To conclude, it was found that the way water is managed is intertwined with political relations, linking it to power structures and control of resources between how water is managed and who influences how it is managed.

7.3.3 Compliance Monitoring and Enforcement (CME)

> There are rules and regulations that we have to follow. We can't just do anything, and we have to stand within the limits of the law. (van Niekerk, personal interview, 2018)

As the name suggests, the role of the CME unit within the DWS is to ensure all those with a water licence or authorisation are compliant, and if not, then enforcement should take place. All Gazettes released have to be monitored for compliance, and this includes the Mine Water Management Unit. This section of the book presents the role of CME and how stakeholders attribute the inconsistencies in CME to issues between government departments and national government giving preferential treatment to municipalities when issues of non-compliance are found, compared to the private sector.

Proactive community members in the Blesbokspruit Forum report many issues of non-compliance to the appropriate government departments such as the CoE, the DMR or the DWS (Blesbokspruit Forum 2019a). These departments are meant to send inspectors out to conduct an investigation and provide reports, but this does not happen (Humphreys, personal interview, 2018; Labuschagne, personal interview, 2018). It is often the case that government officials suggest that they could not identify the reported source of pollution (Storey, personal interview and site visit, 2018) or they do nothing to prevent the pollution and hold those responsible to account (De Jager, personal interview, 2018; Madden, personal interview, 2018), which is highly problematic (Humphreys, personal interview, 2018). Madden and De Jager have physically escorted government officials to show them the non-compliance of coal mines, but nothing is done and it is understood to these key individuals that the government condones illegal activities. A media report substantiated this, accusing the government of "authorising pollution" after water quality tests for the AMD treatment plant exceeded the government's own water quality guidelines (Bega 2017), implying that the CoE is not doing its job or monitoring properly. Community members, such as Adrian Storey, went the extra mile and sought assistance from environmental activist Mariette Liefferink of the FSE to assist with investigations regarding illegal mining. She is a known voice among the community and able to influence government decisions. Even though non-compliance of licences is a result of poor management and lack of care on the government's part, it may seem like these incidents are not severe enough for the government to enforce compliance. Especially after the government is informed by community members.

In an area such as the East Rand where there are numerous industries that need to be monitored, it is impossible for those industries that are non-compliant to get away with it (Humphreys, personal interview, 2018). However, "they [the CoE] know what's happening. They see what's happening, but there is no one to stop it" (Humphreys, personal interview, 2018). Stakeholders raised concerns that only certain industries and government departments are monitored and some get away with non-compliance of their water use. For instance, Barker (personal interview, 2018) suggested that if a private sector company was non-compliant, it would be shut down overnight. However, the same did not apply to local government. He is suggesting that there is preferential treatment. The CME unit has the ability, responsibility and power to revoke the licence if an offender is non-compliant to ensure a reduction on the impact on the environment and for the offenders to take compliance more seriously. However, a CME representative (Klip River Forum 2018) said that support should be given to the local municipalities if found non-compliant to assist the municipality to comply before opening a legal case. This leniency was due to the common knowledge that local and national government do not work together.

Further to the issue of providing municipalities with support, was that the responsibilities of the CME unit are a lengthy process as explained during the Blesbokspruit Forum (2018c). There has to be evidence of non-compliance and attempts to assist the offender to rectify the damage. A criminal case can be opened, but according to section 151 of the NWA, first an investigation, notice, directives, follow-up on this directive has to be completed prior to the execution. This could mean that enforcement is not followed through because of the lengthy process, implying that the "fewer" the problems, the fewer the interventions needed and this requires a change in mindset (van Niekerk, personal interview, 2018). This is evident when it comes to the issue of illegal miners. The DMR has a forum on illegal mining that works closely with the South African Police Service and mines in the area to deal with the problem. However, no one wants to see the prosecution process through (Govender, personal interview, 2018) and therefore, illegal mining continues to take place.

In addition to enforcement being a lengthy process is the lack in staff capacity in the upper levels of the municipal structure, where there are numerous acting heads of departments, and "no one is prepared to make a decision or to give up what they doing. Everyone is protecting himself or herself, and functions are not fulfilled" (Humphreys, personal interview, 2018). Further, there are only 35 enforcement inspectors available in the whole country and no criminal convictions, notwithstanding the fact that 55 out of 111 mines were found to be non-compliant with their water use licence (Liefferink in Blesbokspruit Forum 2018a). In addition to lack of staff capacity, Bosman (personal interview, 2018a) explained that municipalities were the biggest polluters of water due to a lack of funds for maintenance. She emphasised that this is a national government problem, and it should be assisting with funding. The problem is that the government is not held accountable and the public is, and this presents a gap in the law (Bosman, personal interview, 2018a). This creates the perception that the environment is leveraged for socio-economic growth (McKay and Milaras 2017, p. 4).

There is a lack of interest in pollution incidents from the government (Pretorius, personal communication and site visit, 2018), meaning that "the government does not care about the effects on the Blesbokspruit," because known incidents are seldom investigated or resolved (Storey, personal interview and site visit, 2018). An example of power relations that occur between the community and government departments is when it comes to reed maintenance. The community informed the CoE and the GDARD about their concerns with the reeds and are told to not personally maintain the reeds because it is an environmental issue (De Jager, personal interview, 2018). However, when complaints of residential and business owners in the community become extreme, then the departments allow members of the community to assist with maintaining the reeds, because they cannot do it due to a lack of funding. These community members actually have power due to their capacity and access to financial resources to assist with the reed maintenance, which in turn contributes to the management of the Blesbokspruit. However, this case illustrates how the government exerts authoritative power on the community, despite not fulfilling its functions due to a lack of capacity/resources.

The issue of power differences within national government departments surfaced because the non-compliance of coal mines not only impacts on the community, but also on national government (the DWS) operations, such as AMD treatment. For example, in 2019 evidence surfaced of non-compliance of a water use licence from a mining company in close proximity to the AMD treatment plant (Blesbokspruit Forum Google group, 2019). Tlale (personal communication and site visit, 2018b) expressed her concern about mining in the area and concurred that if a mining activity had been authorised by the DMR and the mine was non-compliant, the public might assume the pollution was coming from the AMD treatment plant's operations. This makes the work of the TCTA difficult and expensive because their designs are based on certain quantities of discharge per day to treat the water (Tlale, personal communication and site visit, 2018b). The DMR has the authority to reduce non-compliance by mines and these decisions should take the aim of the AMD STT into account (Joshua, GDARD representative, personal interview, 2019). The AMD plant cost R1 billion and it is unacceptable that departments do not communicate to ensure effective governance (De Jager, personal interview, 2018). This links to the need to enforce legislation and offenders being issued with fines (Labuschagne, personal interview, 2018) and strict decisions about whether the mining venture is worth it (Joshua, GDARD representative, personal interview, 2019). AMD issues could be a result of government and mining companies not acting in the best interests of the environment. Instead of working as partners, they operate as separate entities working solely for their own interests and further enforcing the idea that "political will is lacking to fix the problem" (Erwat 2011, p. 32). This lack of coherence resulted from the complicated relationship between sociopolitical and historical contexts, together with the conflicting and unclarified responsibilities and implementation strategies of the various government departments involved in AMD management. This lack of clarity can be attributed to the power dynamics or relevant lack of power.

The lack of coherence in government fulfilling their responsibilities qualifies the public to question why they should comply with rules when the national government does not (Fourie, personal interview, 2019). "We are acutely aware of these things. There are rules and regulations that we have to follow, we can't just do anything, and we have to stand within the limits of the law" (van Niekerk, personal interview, 2018). This is the reason for AMD treatment being finalised at such a late stage; the government "dragged its feet" (Van der Merwe, personal interview, 2018). The rules are there but not enforced and policy changes must adapt to this (Humphreys, personal interview, 2018). McKay and Milaras (2017, p. 4) find it debatable in the case of the AMD crisis caused by the Grootvlei Mine as to whether the issue could have been avoided had enforcement been in place. They are of the opinion that such urgency is not in government policy. The DMR had not taken any action against the directors of the mine and believed the owners received preferential treatment because of who they were. This implies that the national government has become an agent rather than a controller of the mining industry to serve its own interests of economic growth through exertion of its authoritative power.

The Constitution gives the national government priority in an intergovernmental dispute, but this stands to reason. Local government should also be able to dispute matters, even if it leads to conflict between spheres of government (Humphreys, personal interview, 2018). There has been several attempts and suggestions of the Blesbokspruit Forum (2018c) to have a task team that consists of the South African police, the DMR, the DWS and the CoE to monitor compliance more effectively and for all water users to attend the forum to report back on their water use discharges in the Blesbokspruit. But this would imply that South Africa's sound environmental legislation is lacking strong political will and amended legislation is required to take the inconsistencies into account to tackle environmental liabilities such as AMD.

The lack of enforcement led stakeholders to believe that the government lacked the necessary concern to address ongoing incidents of pollution and non-compliance by, and enforcement actions against, offenders. This influenced social constructions of the water quality. These constructions are linked to power relations because whether action is taken depends on how influential the stakeholder reporting the incident is. It therefore depends on "who knows" and who is heard. It also depends on capacity and resources available for the government to enforce compliance. This is how questions about power and authority arise.

7.3.4 Exemption of an Environmental Impact Assessment for the Use of the Mine Void

EIAs are there because the law cannot cover everything. (Bosman 2018b)

EIAs played a significant role in the research with regard to how the water quality of the Blesbokspruit was socially constructed. Section 5.6.3 of this book presented the perceptions about the significance of an EIA. Stakeholders felt that there was no purpose to the EIA public participation process in which they had participated in several meetings. Further, they felt misinformed about the eventual disposal site and misled due to the AMD STT being implemented under emergency conditions, which was exempted from an EIA. This meant there were unknown consequences for using this site, which created a lack of trust in the government to follow the law. Further, information on the implemented STT process was not communicated to the public and this influenced how stakeholders socially constructed the water quality, based on the possible unknown impacts posed by the STT. This section of the book unpacks why these constructions were linked to the power relations at the national government level.

The common legal understanding of an EIA is that it is a serious examination of the potential effects that a project may have on the environment. It is a legal requirement to ensure the proposed project and activities are environmentally sustainable (National Environment Management Authority 2018). An EIA must be conducted before the project begins. It is considered an offence to commence a project without an EIA approval from the DEA. The DEA is also responsible for the implementation of the EIA. Further, the EIA process requires that the views of those who may be affected by the project are considered, and only then shall an EIA be issued before the commencement of the project (National Environment Management Authority 2018). The community are allowed to request a copy once the EIA is signed and issued by the DEA, and if any of their concerns were not included in the final EIA, they are allowed to dispute it (Bosman, personal interview, 2018a).

As regards the EIA for the sludge disposal site, Maurizi[8] (personal interview, 2018) claimed that the public received letters to say the AMD STT project had been

[8] A. Maurizi is a water quality manager at the CoE and a Springs resident.

approved and was continuing. This made it difficult to get additional information, especially as to whether the public's comments were considered or whether there were further amendments and if in practice the EIA was adhered to (Maurizi, personal interview, 2018). When the project was implemented, the public became aware through various channels (e.g. other community members and the Blesbokspruit Forum) that a new sludge disposal site was being used. An important finding is the interaction between the influences on the social constructions. Use of the mine void as the sludge disposal site instead of the EIA-approved site was based on a legal justification (and therefore legal power), but still it produced critical constructions. This is how the idea raised by key individuals about AMD STT being attributed to authoritative and legal power from national government came about.

Potential impacts to surface and groundwater resources were identified and assessed as part of the EIA process conducted by Digby Wells, and suitable mitigation, management and monitoring requirements were stipulated in the report as a condition for approval. These requirements meant reducing the potential impacts on the water resource to meet the NWA requirements (Digby Wells Environmental 2014). However, the EIA-approved site was changed and the mine void was implemented under emergency measures, with the exemption for an EIA granted. This created trust issues between the public and national government because the purpose of an EIA could not be fulfilled. Fourie[9] (personal interview, 2019) suggested that an EIA is a tick box exercise, and unless there is a clear "no-go", public concerns would not influence the process. This leads to public frustration and the opinion that national government believes that it is not bound by an EIA (De Jager, personal interview, 2018; Madden, personal interview, 2018; Maurizi, personal interview, 2018). Maurizi suggested that this was a general problem, the public should be involved and everyone should be bound by the recommendations in that EIA because its purpose was to ensure approved processes are followed.

When the DWS realised that the Grootvlei Site 6/L/16 was no longer available, the AMD project team had to find out whether a new EIA was required and what process to follow (Tlale, personal interview, 2018a). The AMD plant was already constructed, and to continue the project, the team requested the necessary permission and were given a set of conditions that made the process more palatable (Govender, personal interview, 2018). The TCTA had to identify possible inconsistencies regarding the AMD project. Once they received the support, advice and approval from the DWS where needed, consultations took place with the DEA, the DMR, the DWS and the TCTA on a way forward. As a result, the DEA could not pinpoint an activity that would be triggered by the TCTA's use of the mine void. Nobody knew exactly what the process was, but "the clock was ticking" (Tlale, personal interview, 2018a). The main purpose was to ensure that AMD water was pumped before reaching the environmental critical level (ECL). At this stage, the project became an emergency, because the treatment was ready to be implemented but could not begin without a disposal site for the sludge. The DWS had to make a

[9] B. Fourie is an aquatic ecologist.

decision and issued the TCTA with a directive to go ahead with the project on the condition that there would be frequent monitoring to ensure that the project ran accordingly, as recommended by the DEA (Govender, personal interview, 2018; Tlale, personal interview, 2018a). The Water and Sanitation Minister permitted the implementation of the project as an "emergency works water management" in terms of section 110(2) of the NWA and a directive was issued by the DEA to implement the project (TCTA 2019). Govender (personal interview, 2018) said that he battled to make people appreciate that because the project was commissioned by virtue of section 110(2), this excluded the DWS from conducting an EIA (RSA 1998a, p. 51).

Springs resident, Madden (personal interview, 2018), suggested that the emergency status became attributed to the DWS being forced into it and because the DWS had no other interest than addressing the "emergency". Bosman (personal interview, 2018a) and Maurizi (personal interview, 2018) questioned on what grounds a project could be regarded as an emergency and for whom it was an emergency. AMD was a problem in 2002 and named a crisis back then. It was now over 16 years later (at time of interview), "so what is the emergency now when it was a crisis that many years ago" (Bosman, personal interview, 2018a). "We need to look at the reality instead of using the idea of a threat" (Bosman, personal interview, 2018a). Govender suggested that an EIA was a misnomer, when an EIA is conducted it means that the DEA had seen and approved it, which meant that it had taken a decision for a specific activity and site. However, what the public is unaware of is that the project is not compelled to proceed along approved lines, because there is a possibility that the project will not run its course (Fourie, personal interview, 2019; Govender, personal interview, 2018). A project is compelled to receive authorisation for the new activity or a deviation from the activity, and this was done when the decision to use the mine void for the disposal of sludge was taken (Govender, personal interview, 2018). If a plan is made publicly available, it must be workable and if not, then there must be an additional plan in place that has been further investigated (Maurizi, personal interview, 2018). The mine void became a feasible option, according to Environmental Manager at the TCTA, Tlale (personal interview 2018a), only once it became known to the DWS and the TCTA that there was a conflict with the use of the initial sludge disposal site. The fact that the public suggested the use of the mine void during the EIA process for the Grootvlei Site 6/L/16 tailings storage facility (TSF), but the suggestion was regarded by the project team as being flawed due to its potential impact on groundwater, would point towards the DWS admitting that it was not investigated at that stage. This would mean that the national government is indirectly admitting that there is no purpose for the EIA public participation process (which involves investigating other suggestions made). Further, there were cost and time implications. This is why problems with water governance are often linked to power relations, where the society is viewed as an obstacle to the DWS achieving "good water governance". Therefore, information on changes made is not publicly shared.

For the public, immense effort from their part was put into the EIA process, and this is why it created a lot of negativity for the community, the CoE and other local authorities when the process was not followed (Maurizi, personal interview, 2018).

This indicated that stakeholder engagement is a crucial part of water governance. Citizens have the right to be represented and heard (Barker, personal interview, 2018). However, this depends on who is working on the project and who determines what decisions are made and what aspects are focused on, and therefore, the main concerns are not assessed from a broader perspective (Fourie, personal interview, 2019; Maurizi, personal interview, 2018).

There was also a perception of lack of objectivity on the part of the environmental consultants, Digby Wells, despite them carrying a lot of weight (or having reputational power) in the industry (Barker, personal interview, 2018; Pillay, personal interview, 2018). Even though they had an institutional obligation to conduct an EIA on behalf of the DWS, they also had an obligation to take the public participation process seriously (Barker, personal interview, 2018). In Pillay's[10] view (personal interview, 2019) the environmental consultant only conducts the research and the client [DWS] makes the final decision. Fourie (personal interview, 2019) explained that more than likely the suggestions put forward during the EIA public participation process are approved by a less qualified person and the recommendation is sometimes completely opposite to what was suggested. Such a situation arises because an unqualified person is in an authoritative position, meaning the expert is not making the final decision but the least experienced person has the most decision-making power. This results in changes to the whole project because the person making the decisions does not understand the situation from an expert point of view (Fourie, personal interview, 2019). This meant that a high-level strategic understanding of reality in the DWS, the DEA or at any political level was lacking and the people who did understand were pursuing their own interests which did not always involve advising their political heads (Bosman, personal interview, 2018a). "AMD is an opportunity to make money", meaning even if there is an understanding, the government is going to first prioritise how money can be made (Bosman, personal interview, 2018a). Bosman's view can be justified by the fact that the CoE was not informed of decisions regarding AMD treatment that fall under its jurisdiction but will have to take responsibility for monitoring the quality of the water (Fourie, personal interview, 2019). However, if the TCTA does something wrong, nothing will happen because it is employed by the DWS, and Fourie is of the view that this is what happened with the exemption for the EIA, implying that "something went wrong". A lack of communication and information to the public resulted in stakeholders speculating what occurred, leading to the view that the national government made decisions by exploiting its authoritative or legal power.

Even though the situation would have been worse if nothing had been done, one cannot accept this process continuing indefinitely (Maurizi, personal interview, 2018). Once it is no longer an emergency, what next? Will timeframes be adhered to in the future, or will the same problem occur, where targets are not reached (Maurizi, personal interview, 2018). Fourie (personal interview, 2019) suggested that because the STT was implemented as an emergency, the actual treatment was the main focus

[10] M. Pillay is an environmental consultant at Digby Wells.

rather than incorporating all risks such as the possible socio-economic aspects. However, no options existed from a legal, economic, social and environmental perspective. The project was executed on a site-by-site basis (Bosman, personal interview, 2018a). This is why one gains the impression that an EIA is insignificant. However, Maurizi (personal interview, 2018) suggested that it could be worse; "the public have to be convinced that the emergency measure was the best option". Even though it is not an easy decision to make: "I wouldn't want to be in the shoes of those who have to make such decisions. Therefore, it's not easy to point a finger" (Maurizi, personal interview, 2018). Maurizi has gained power in terms of her credibility and the strong reputation she has built in her role as the CoE Water Quality Manager: she is referred to as the "go-to person" (Van der Merwe, personal interview, 2018). She contested the EIA process, but at the same time acknowledged the difficulties that the DWS could have experienced.

Public frustration comes from the large sums of money that went into completing the discarded EIA, and furthermore, from the public's exclusion from the implemented STT process. This exclusion led to heightened public emotions regarding the STT process which exemplified the views of the actual treatment. This finding indicated that the public feeling is that decisions made were due to poor planning. Poor planning led to non-adherence to the public participation during the EIA process, which meant authoritative power was used to enforce decisions. The public may actually have understood what an "emergency measure" entailed, but the lack of communication and information sharing led to a lack of trust in the government, and this influenced individuals' constructions of the water quality.

7.4 Communication

> Projects can fail because communication about potential risks are not clear. (Bosman 2018b)

Even though the public were involved in the EIA process for the Grootvlei Site 6/L/16 TSF, they were not communicated with regarding the changes made when the STT of AMD was implemented. Projects are more successful when all stakeholders who are impacted become involved through their participation, because then they will be made aware of the potential risks of the project (Bosman 2018b). In evaluating risks of a project, the expectations of society are revealed, and this depends on sociopolitical input. In the case of AMD treatment, both the public and national governments have an interest. The government has a responsibility to reduce the risk, but this risk is reduced through public taxes, about which people should be informed. The more efficient the management is, through communication and information sharing, the lower the level of risk will be (Bosman, personal interview, 2018a). Access to information is a right, according to Chapter 2, section 32 of the Constitution (RSA 1996, p. 13). Everyone has the right of access to any information held by the State and any information that is held by another person and that is required for the exercise of protection of any rights. The community does a lot to get involved (Humphreys, personal interview, 2018) and influential members of the community inform other community members, whereas it should be the government's role to communicate with the public to ensure credible information is transferred. However, this was not the case regarding the AMD treatment.

Communication and information played a contributory role in how key individuals and stakeholders socially construct the water quality of the Blesbokspruit (discussed in Sect. 6.6). In this chapter communication and information are discussed in terms of how they are linked to power. First, those individuals who have access to information can use effective communication as a form of power, compared to those with unreliable or no information, who have less power. Second, communication coming from those in authoritative positions will display more power than that of other stakeholders, because the communication of authoritative institutions is normally regarded a more credible and influential source of communication. Third, those stakeholders who generally communicate more effectively due to their knowledge have more impact or influence on the perceptions of others than those who communicate poorly or are not as knowledgeable. Such stakeholders (e.g. influential community members) become credible sources of information, and they become trusted by others even if they do not hold senior positions in managing the Blesbokspruit.

Communication and information on AMD treatment were transferred by various stakeholders through the means mentioned above. Owing to the STT being implemented as an emergency measure, no public information was available, communication was not clear and no risk assessments and potential impacts were provided. When there are potential increased levels of pollution, the response depends on what is acceptable to the public and it becomes political in nature (Bosman 2018b). The public did not have an opinion on the actual implementation of the AMD STT, because the DWS restricted communication and information to avoid further issues and negativity. The aim of this section is to present the power dynamic between individuals and institutions through their interactions to justify how communication influences social constructions of water quality and are embedded in power relations, how those in authoritative positions exert their power on other stakeholders and how these stakeholders compete with such power dynamics. The existence of power dynamics is explained through the miscommunication about the sludge disposal site and the LTT.

7.4.1 Miscommunication About the Sludge Disposal Site

> Corruption suppresses any form of development, its bad news. (van Niekerk, personal interview, 2018)

The feelings about the sludge disposal site came from stakeholders feeling misinformed and misled on how the decision to use the mine void was made. This suggested that the public did not influence government decisions even though water governance should include their views through public participation. How key individuals and stakeholders socially constructed the water quality was linked to power relations and influenced how communication about the sludge disposal took place, which defined their role and value (discussed in this section).

The use of the mine void was suggested by interested and affected parties (IAPs) during the EIA process for the sludge disposal site, because at that stage many IAPs were objecting to the Grootvlei Site 6/L/16 due to it being an above-ground site and therefore a higher pollution risk. However, the mine void was regarded as flawed because it could potentially impact on groundwater. This is where the negative perceptions of the mine void surfaced. When the STT was launched and stakeholders became aware through various means that the mine void was used due to emergency measures, concerns regarding potential impacts on groundwater and increased levels of water pollution were raised. This can mean that as knowledge of science increases, people adapt to what is acceptable and what is not, and this too can change. Therefore, what is regarded as pollution can change to what is acceptable and what is not (Bosman, personal interview, 2018a). There are different levels of acceptability for different people. If it affects their lives, then it tends to be viewed as unacceptable and the pollution risk is viewed as high. If it does not cause direct risks, then the pollution levels are low (such as agriculture's need for quantity of water more than quality of water). The levels of acceptability regarding the use of the mine void stemmed from power relations between the national government and other stakeholders, based on how decisions were made and how communication took place.

The lack of communication regarding the use of the mine void for the sludge disposal is twofold. First, the public claimed that they were not aware that the sludge disposal site had changed from the EIA-approved Grootvlei Site 6/L/16 to the mine void (Madden, personal interview, 2018; Maurizi, personal interview, 2018; Pillay, personal interview, 2018). Second, the public claimed that they were not made aware of why the sludge was disposed of in the mine void (Madden, personal interview, 2018).

As regards the awareness of the disposal site, Maurizi (personal interview, 2018) believed that if people heard information by accident, it causes public frustration. This is because it is viewed as a waste of their time, as in the case of the mine void being used for which they had already provided their input during the EIA process and were not communicated with. The first time Madden (personal interview, 2018) heard of the mine void being used was through the local paper, the *Springs Advertiser*, after the launch had taken place: "our wonderful Minister of Water shouts out loud, oh, we're putting all the stuff down the mine void" (Madden, personal interview, 2018). Digby Wells, who conducted the EIA for the sludge disposal site, only realised that the site had changed when they were taken on a tour of the plant during the launch (Pillay, personal interview, 2018). Mashau, GDARD representative (personal interview, 2019) said: "we did not know where the sludge was being deposited and heard this during the Blesbokspruit Forum". Thus, key individuals from the community, provincial government and the environmental consultants indicated that they were uninformed.

As regards the lack of understanding as to why the mine void was used, the public reached different conclusions (Barker, personal interview, 2018; Bosman, personal interview, 2018a; van Niekerk, personal interview, 2018). Van Niekerk believed that it was because the project team had changed their mind and did not want to inform the public, "corruption suppresses any form of development. It's bad news" (Van Niekerk, personal interview, 2018). This implies that when information is withheld from the public they view it as being due to hidden agendas or because of unlawful activities taking place.

The idea existed that the disposal of sludge was illegal because a separate EIA should have been conducted for the discharge of neutralised water and the new sludge disposal site (Bosman, personal interview, 2018a). Barker (personal interview, 2018) suggested that someone was benefitting by allowing such decisions to be made. Based on these views of these key individuals, communication about the use of the mine void was not transparent and linked to power stemming from ineffective communication and authority. Govender (personal interview, 2018) from the DWS suggested that

> specifically regarding Gauteng AMD challenges, anyone who is of the opinion that there is no comunication is seriously misled or misinformed. We engaged with the DWS Gauteng [provincial office] and the national office has been invited to the forums, specifically where AMD is of interest and there is always a representative present.

According to him, there is a competency in the provincial office and the national office that deals with mine water management and always ensures representation at the forums. Therefore, Govender (personal interview, 2018) strongly suggested that there has been communication on AMD STT, because the TCTA is present at these catchment forums to fulfil the communication function.

In February 2017 an extensive update on AMD STT was presented at the Blesbokspruit Forum by Tlale of the TCTA (Blesbokspruit Forum 2017a). The update included the issues pertaining to the water use licence and that the project was implemented as a government waterworks, which exempted the need for environmental authorisation, and, by implication, the need for a water use licence. Further, Tlale (Blesbokspruit Forum 2017a) communicated in the forum that the sludge was disposed into the mine void as a pilot project though no EIA had been conducted and that the LTT would be integrated into the STT. She further explained why the site had changed and provided information on how deep down the void the sludge was being pumped, indicating that there was a minimal risk of additional pollution and groundwater was not pumped so there would be no impacts on the groundwater Table. A review of the forum's minutes for this research indicated that information pertaining to the sludge disposal site was disclosed at this meeting to the 38 stakeholders who were present. They also had an opportunity to ask questions. The Blesbokspruit Forum is a public forum, but this information was disclosed only after a representative of Rand Water had questioned the AMD treatment operations. Further, this was the only platform where the DWS communicated about the use of the mine void. Therefore, only specific people gained access to this information, such as those who attended this meeting and those who would read about it in the *Springs Advertiser*, where the media alerted the public to the sludge disposal site after the launch of the treatment plant took place.

Certain stakeholders who are directly impacted by the disposal, such as the CoE and GDARD, were only informed through these means, and this created negative perceptions of the process due to restricted communication and information sharing. These departments play a direct role in managing the water quality of the Blesbokspruit and were not informed initially about the changes in the STT process. Even though Mashau stated, "we were so excited about the treatment plant, we could have forgotten to ask those questions" (GDARD representative, personal interview, 2019), Fourie (personal interview, 2019) suggested that when information is not shared openly, rumours spread, and incorrect information is transferred; it is due to a lack of transparency.

Turton's[11] (personal communication, 2018) explanation unpacks the possibility as to why communication from the AMD project team could have been weak. Regulations do not allow for deposition into the void, and it is almost impossible to implement due to the natural movement of water of surface tailings dams into the void as part of the hydrological cycle. In the western basin, the practice of discharging barren (depleted) tailings into the West Wits Pits (the practice in this basin) had beneficial effects that no mining activist would make public knowledge (Turton, personal communication, 2018). This could mean that the project team could also have been fearful of being open about this process due to the negative comments that could arise from the public due to differences of opinion over the use of the site. In addition, there were many unknowns in using the mine void, which meant questions could not be answered. This idea is linked to Tlale's explanation that the mine void was not researched or used before, but if proven successful, it would form a good solution in Gauteng or elsewhere. It would mean that all sludge from AMD treatment could be disposed of underground and more land would be available and accessible for other uses, and the problem with mine dumps could be minimised or resolved.[12] This is why the mine void as the alternative site was considered the most feasible option in comparison to other options (Tlale, personal interview, 2018a). Due to the public not having this information, a lack of understanding arose which led to some individuals believing that the use of the mine void was illegal. According to Govender (personal interview, 2018), it is not illegal to dispose of the sludge in the mine void. The DWS had consulted with other relevant government departments before disposing of the sludge and received authorisation (as discussed in Sect. 7.3.4). "What needs to be understood is that the idea that it's illegal is a perception" (Govender, personal interview, 2018). This perception, however, stemmed from the lack of, or incomplete, communication. Even though the DWS had the authorisation to dispose in the mine void, the DWS would not tell the public exactly what they were doing because "the truth hurts" (Barker, personal interview, 2018). This means that the DWS would have to provide answers that they possibly had no reasoning

[11] A. Turton is a water specialist.

[12] The sludge would have been disposed of at a TSF, which is the case in the other basins.

for. "When something becomes politically sensitive that's when it falls through the cracks, and activists like Mariette Liefferink become crucial because they are prepared to dig and look for answers, and we need these kinds of people" (Barker, personal interview, 2018).

Govender (personal interview, 2018) explained that the problem with communication was that as a government official he was the only institutional link regarding the AMD project, and the TCTA representative had an obligation on behalf of the DWS to provide clarity during the forums. Once people are aware of what they are doing, they will be more positive towards the process (Tlale, personal interview, 2018a). Govender (personal interview, 2018) acknowledged that it is important to list the issue of clarity – or lack thereof – as a finding of the process and going forward provide clarity on these issues through the forums and through media statements but the effectiveness of this communication channel in reaching the general public is always dubious. While there is a lack in communication from the national government, information is communicated and transferred through other credible sources who are not necessarily in authoritative positions but have influential power, such as Mariette Liefferink. Further, there is a lack of staff capacity, which possibly contributed to why information is not shared, because there are only a few people directly involved in the AMD project. This indicated how legal power is used to make decisions due to a lack of capacity.

Even though communication about the mine void was attributed to government's authoritative power, key individuals from the DWS and the TCTA acknowledged that their communication was weak. This could mean that there was a level of concern to openly discuss the issues and inform the public. It could be to avoid possible questions on the social and environmental impacts of using the mine void that the project team could not answer, because it is a pilot project. If the decision to use the mine void was communicated to the public before they found out about it through other means, their receptivity to the issue could have been stronger. However, for these possible reasons, decisions are made by the government from an authoritative position rather than from a position of inclusivity. Poor communication about decisions made influenced how individuals socially construct the water quality of the Blesbokspruit. This led the public to believe that the decision to use the mine void was based on an act of power rather than what was best for the environment and the people.

7.4.2 Long-Term Treatment: Issues and Conflict

> The long-term treatment is becoming longer. (Barker, personal interview, 2018)

Discussions around the LTT are conflicting in nature due to the time it has taken to implement the long-term solution. Therefore, by not informing the public, delays in the process have led to concerns for stakeholders about when the LTT would take place and further if it would be implemented in the same manner as the STT.

In 2014, during the EIA process for the sludge disposal for the STT, stakeholders were informed that the STT would continue for 5–8 years and then the LTT would commence (Digby Wells Environmental 2014). In February 2018 members of the Blesbokspruit Forum were informed by the TCTA that the LTT was on the way and that the EIA process had commenced (Blesbokspruit Forum 2018a). Nemai Consulting was the appointed environmental consultant. However, the engineering professional service provider who would carry out the long-term solution had not been appointed and consultations still had to take place with the DEA (Blesbokspruit Forum 2018b). By August 2018 the engineering provider had still not been appointed, due to a lack of funding (Blesbokspruit Forum 2018c). Further, the decision about who should handle the issues regarding the EIA application process between the DEA and the DMR was not clarified (Blesbokspruit Forum 2018c).

Information about the EIA for the LTT was not disclosed to the public. Interested parties were added to the database to be informed of the public participation process but the environmental consultants were not given authorisation by the AMD project team to directly discuss information with members of the public (Govender, Tlale and Naidoo, personal communication, 2018). This hints at the political nature of LTT. Stakeholders were frustrated and negative about the LTT possibly because they were not informed and updated about when the treatment would commence since the implementation date kept changing.

> Several years down the line and we're still in design phase. It seems as if the short-term intervention is becoming permanent rather than temporary. This means we are getting no returns on this process, whereas if there was funding to implement the long-term, money could be made [potable water sold to an end user] out of that process. (Barker in Blesbokspruit Forum 2018c)

The knowledge that stakeholders had of the EIA process for the sludge disposal site, taking almost 2 years to complete and then not being relevant as another site was used, led to concerns about whether the LTT would also be implemented under emergency measures to justify the exclusion of an EIA since it took so long to secure funding (Blesbokspruit Forum 2018c). In August 2020, Liefferink (in Blesbokspruit Forum 2020b) provided feedback to the forum on the LTT progress made to date. R990 million was needed for the LTT to commence but funding remained the main obstacle (Blesbokspruit Forum 2020b). Stakeholders were concerned that this could mean that legal power could possibly be used by the DWS to exempt them from conducting an EIA for the LTT due to lack of financial resources.

There are already mixed messages and limited communication that exist regarding the LTT, and the time factor creates uncertainty and lack of trust in the government's commitment to implement this treatment. "There were promises that were made with regard to the short-term solution but it seemed as though the short-term solution was rapidly becoming the long-term solution and the TCTA needs to do something about it" (Blesbokspruit Forum 2018c). This created further negativity around the STT due to the possibility that it continues indefinitely and therefore continue to discharge highly saline water into the Blesbokspruit and eventually impact on the Vaal River system (Blesbokspruit Forum 2020b; Bosman, personal interview, 2018a; Liefferink, personal interview, 2016). "[Therefore], the public needs to know there is commitment" (Barker in Blesbokspruit Forum 2018c).

In addition to concerns about the impact on the Vaal catchment, stakeholders claimed that a considerable amount of water was currently being partially treated that could potentially be reused or put back into the system, but it is seen as impossible because the long-term solution has not been finalised. Tlale explained that the treated water would most likely be sold to surrounding users (Blesbokspruit Forum 2017b). The sludge will still be produced and managed during the LTT, but instead of partially treated/neutralised water, the water will be of potable standards (Pillay, personal interview, 2018). The desalinated/potable water will go to an end user who is yet to be determined (Govender, personal interview, 2018).

Pillay (personal interview, 2018) said that there were many differences over how the water should be distributed: Rand Water would want to sell it; the farmers and miners would want it to be put back into the river and the environmental sector feels that if it is clean water it cannot be wasted by putting it into a polluted river, but at the same time, there needs to be a flow in the river. There is not enough water; therefore it would be ideal to "put it in the taps" (Maurizi, personal interview, 2018), but the mindsets of people regarding water quality are different depending on their needs. The cost incurred for the LTT would not make economic or environmental sense to put clean water back into the river. This implied that there will be immense conflict when the LTT commences, because it will be funded by the taxpayer (Pillay, personal interview, 2018). Further, when the LTT will commence and how the water will be distributed or used has not been finalised. This implies that even though the taxpayer is contributing to treating the water, the final decision on what will be done with the water will be made by the government, who has the responsibility to treat the water.

Communication by the national government to the public regarding the LTT is still very vague which influences social constructions of the water quality. This demonstrated that how communication is shared is linked to power. Representatives of the forum showed appreciation for the information shared by Mariette Liefferink but felt that it was not civil society's responsibility to relay information but that of the national government:

> We find information from various other sources instead of the direct source, sharing information is such an important aspect to keep in the forums. The community must be seen as a resource and not the enemy. The community wants to help, but the government needs to be open and transparent, we are in this boat together and should all be allowed to contribute. (Representatives of the Blesbokspruit Forum 2020b)

The discussion in this section demonstrated how influential stakeholders, such as Mariette Liefferink who have access to information, used it to effectively communicate as a form of power: "civil society must play a watchdog role", she said (in Blesbokspruit Forum 2020b). Liefferink thus becomes increasingly known as a credible source of information because of her ability to communicate awareness to the public when the government does not. Even though other stakeholders have access to information through influential stakeholders, such as Liefferink, they would prefer that those in authoritative positions, such as the government, communicate directly with them. Stakeholders who are less influential may feel this way because it is their constitutional right to have access to information rather than it being relayed through other means, even if the communication comes from someone who is credible and reputable. Further, the purpose of the Blesbokspruit Forum is for communication and information to be shared among all stakeholders involved in managing the catchment. In addition, the fact that the relevant government departments are represented at the forum but are not the ones sharing the information could imply lack of communication between spheres of government. Lack of communication therefore strongly influences how water quality is socially constructed, depending on what is known about how the water is managed.

7.5 Conclusion

This chapter explained how power relations play a crucial role in how people socially construct environmental realities. Differences in opinion over AMD STT indicated that decision-making is taken by national government, even though the responsibility for water management was shared between spheres of government. This implies that co-operative government as stipulated in the Constitution is far from being reached, even though national government acknowledged that clarity was needed between spheres of government. This led to issues with how the water is managed due to lack of compliance and enforcement measures. The community assists in this regard and are known to raise issues. Some community members are known to be influential and have assisted with how decisions are made or with how information is accessed and then shared with the public. Water governance in South Africa is seen as a highly political issue, fuelled by uncertain roles and responsibilities between spheres of government. Therefore, water struggles between stakeholders regarding water uses influence how constructions are formed and stem from existing power relations. Conditions for water governance, such as inclusion of the public, transparency in communication and information sharing, would need to be considered for improved management of water to ensure the same power struggles with the STT are not repeated during the LTT of AMD.

References

Barker, A. 2018. Consultant: DRD Gold. Personal interview, 4 June. Panorama, South Africa.

Bega, S. 2017. Residents left in dark over acid mine drainage treatment. *IOL News,* 11 February. Available at: http://www.iol.co.za/news/south-africa/gauteng/residents-left-in-dark-over-acid-mine-drainage-treatment-7709811. Accessed on: 15 February 2017.

Blesbokspruit Forum. 2017a. Minutes Blesbokspruit Forum, 2 February. Available at: http://www.reservoir.co.za/forums/vaalbarrage/blesbok_forum/blesbok_home.htm. Accessed on: 31 January 2020.

Blesbokspruit Forum. 2017b. Minutes Blesbokspruit Forum, 4 May. Available at: http://www.reservoir.co.za/forums/vaalbarrage/blesbok_forum/blesbok_minutes/archive/BF_Minutes_04May2017.pdf Accessed on: 31 January 2020.

Blesbokspruit Forum. 2018a. Minutes of Blesbokspruit Forum meeting and personal attendance, 8 February 2018.

Blesbokspruit Forum. 2018b. Minutes of Blesbokspruit Forum meeting and personal attendance, 11 May 2018.

Blesbokspruit Forum. 2018c. Minutes of Blesbokspruit forum meeting and personal attendance, 8 August 2018.

Blesbokspruit Forum. 2018d. Minutes of Blesbokspruit Forum meeting and personal attendance, 8 November 2018.

Blesbokspruit Forum. 2019a. Minutes of Blesbokspruit Forum meeting and personal attendance, 7 February 2019.

Blesbokspruit Forum. 2019b. Minutes of Blesbokspruit Forum meeting, 9 May 2019.

Blesbokspruit Forum. 2019c. Minutes of Blesbokspruit Forum meeting, 8 August 2019.

Blesbokspruit Forum. 2019d. Minutes of Blesbokspruit Forum meeting and personal attendance, 7 November 2019.

Blesbokspruit Forum. 2020a. Minutes of Blesbokspruit Forum meeting, 6 February 2020.

Blesbokspruit Forum. 2020b. Minutes of Blesbokspruit Forum meeting and virtual attendance via Zoom platform, 6 August 2020.

Blesbokspruit Forum Google group. 2019. Notification of collapse of an embankment. E-mail communication, 14 September 2018.

Bosman, C. 2018a. Consultant for Environmental health services. Personal interview, 12 April. Pretoria, South Africa.

Bosman, C. 2018b. IWRM, the NWA and Water Use Authorisations: Focussing on WULAs, and IWWMPs. 13 – 15 March 2018. Carin Bosman Sustainable Solutions (CBSS) Training Course: Pretoria.

Business Tech. 2017. SA Water department is broke. *Business Tech*, 12 February. Available at: https://businesstech.co.za/news/finance/156936/sa-water-department-is-broke-report/. Accessed on: 25 February 2020.

Creamer, M. 2019. Menar's Canyon offers Eskom cheaper coal on open-book basis. *Mining Weekly*, 19 March. Available at: http://www.miningweekly.com/article/menars-canyon-offers-eskom-cheaper-coal-on-open-book-basis-2019-03-19. Accessed on: 4 April 2019.

De Jager, P. 2018. Lawyer and Springs resident. Personal interview, 25 January. Springs, South Africa.

Department of Water and Sanitation (DWS). 2017. *Mine Water Management: Policy Position Draft for External Consultation and Discussion*. Gazette No. 658. Available at: https://www.greengazette.co.za/notices/national-water-act-36-1998-mine-water-management-policy-position-draft-for-external-consultation-and-discussion_20170707-GGN-40966-00658.pdf. Accessed on: 18 October 2018.

Department of Water and Sanitation (DWS). 2018. *National Water and Sanitation Master Plan. Valuing Water Dignifying Sanitation*. Presentation to Provincial Consultation Workshop Preparation for the Water and Sanitation Operation Phakisa DWS: RSA.

Department of Water and Sanitation (DWS). 2019. *National Water and Sanitation Master Plan. Volume 1: Call to Action, Ready for the Future and Ahead of the Curve*. Pretoria: DWS.

Digby Wells Environmental. 2014. *Environmental impact assessment for the construction of the proposed sludge disposal facility and pipeline associated with the treatment of acid mine drainage from the eastern basin of the Witwatersrand, Gauteng: Full comment and response report*. Johannesburg: Digby Wells and Associates.

Erwat, TI. 2011. *Acid mine drainage in the Gauteng Province of South Africa: A phenomenological study on the degree between stakeholders concerning a sustainable solution to acid mine drainage*. Master's dissertation. Stellenbosch: University of Stellenbosch. Available at: http://scholar.sun.ac.za/handle/10019.1/17978. Accessed on: 1 June 2020.

Fourie, B. 2019. Aquatic Ecologist and Consultant: City of Ekurhuleni (CoE). Personal interview, 8 February. Midrand, South Africa.

Federation for a Sustainable Environment (FSE). 2019. Objection to the water use licence application/authorisation in terms of the integrated environmental authorisation for the Palmietkuilen mining project near springs, Gauteng. Blesbokspruit Forum. E-mail communication, 9 May 2019.

Gauteng Wetland Forum. 2018a. Minutes of Gauteng Wetland Forum meeting and personal attendance, 23 February 2018.

Gauteng Wetland Forum. 2018b. Minutes of Gauteng Wetland Forum meeting and personal attendance, 25 May 2018.

GDARD Representatives (Jonhasi, C. Maluleke, R., Joshua, Q. & Mashau, B.) 2019. Gauteng Department of Agriculture and Rural Development (GDARD). Personal interview, 5 March. Marievale Bird Sanctuary, South Africa.

Govender, B. 2018. Chief directorate Mine water management unit: Department of Water and Sanitation. Personal interview, 1 March. Pretoria, South Africa.

Govender, B. Tlale, S. & Naidoo, N. 2018. AMD treatment plant EIAs. Personal communication, 12 April 2018.

Hoekstra, A., Buurman, J. & van Ginkel, KCH. 2018. Urban water security: A review. *Environmental Research Letters*, (13):1–15. Accessed on: 23 May 2018. Available at: https://www.researchgate.net/profile/Arjen_Hoekstra2.

Humphreys, J. 2018. Ekurhuleni Ward Councillor: Bedfordview. Personal interview, 11 April. Nigel, South Africa.

Klip River Forum. 2018. Minutes of Klip River Forum meeting and personal attendance, 8 May 2018.

Labuschagne, W. 2018. Ekurhuleni Ward Councillor: Nigel. Personal interview, 11 April. Nigel, South Africa.

Liefferink, M. 2016. CEO: Federation for a Sustainable Environment. Personal interview, 21 November. Bryanston, South Africa.

Madden, S. 2018. Environmentalist and Springs resident. Personal interview, 8 February. Springs, South Africa.

Maurizi, A. 2018. Manager: Water Quality Division, Ekurhuleni. Personal interview, 3 May. Springs, South Africa.

McKay, TJM. & Milaras, M. 2017. Public lies, private and the forced closure of Grootvlei Mine, South Africa. *The Journal for Transdisciplinary Research in Southern Africa*, 12(1). Available at: https://td-sa.net/index.php/td/article/view/347/399. Accessed on: 12 June 2019.

Mouton, S. 2018. From crisis to opportunity: Lessons from Cape Town water shortage. Available at: https://www.wits.ac.za/news/latest-news/research-news/2018/2018-05/from-crisis-to-opportunity.html. Accessed on: 28 May 2018.

Mpofu, C., Morodi, TJ. & Hattingh, JP. 2018. Governance and socio-political issues in management of acid mine drainage in South Africa. *Water Policy*, 20:77–89. Available at: https://iwaponline.com/wp/article/20/1/77/38136/Governance-and-socio-political-issues-in. Accessed on: 1 June 2020.

National Environment Management Authority. 2018. *Environmental regulations*. South Africa: National Environment Management Authority. Available at: https://www.nema.go.ke/index.php?option=com_content&view=article&id=27&Itemid=167. Accessed on: 1 November 2018.

Oelofse, S. 2010. *Spheres of SA government, responsibilities and delivery*. Pretoria: Waste and Society, CSIR. https://researchspace.csir.co.za/dspace/bitstream/handle/10204/4760/Oelofse6_2010.pdf?sequence=1&isAllowed=y. Accessed on: 21 April 2020.

Pillay, M. 2018. Business Development Executive: Digby Wells Environmental. Personal interview, 16 January. Bryanston, South Africa.

Republic of South Africa (RSA). 1996. *The Constitution of the Republic of South Africa*. Pretoria: Government Printer.

Republic of South Africa (RSA). 1998a. *National Water Act, no 36 of 1998*. Pretoria: Government Printer. Available at: http://www.dwaf.gov.za/Documents/Legislature/nw_act/NWA.pdf. Accessed on: 14 April 2013.

Republic of South Africa (RSA). 1998b. *National Environmental Management Act no 107 of 1998*. Pretoria: Government Printer. Available at: https://www.gov.za/documents/national-environmental-management-act. Accessed on: 14 April 2013.

Republic of South Africa (RSA). 2002. *Mineral and Petroleum Resources Development Act no 28 of 2002*. Pretoria: Government Printer. Available at: http://www.energy.gov.za/files/esources/pdfs/energy/liquidfuels/act28r.pdf. Accessed on: 14 April 2013.

Republic of South Africa (RSA). 2008. *National Environmental Management: Waste Act no 59 of 2008*. Pretoria: Government Printer. Available at: https://www.gov.za/sites/default/files/gcis_document/201409/35572gen615.pdf. Accessed on: 10 March 2020.

Storey, A. 2018. Nigel resident. Personal interview and site visit, 26 January. Nigel, South Africa.

Tlale, S. 2018a. AMD project manager: TCTA. Personal interview, 9 March. Centurion, South Africa.

Tlale, S. 2018b. AMD project manager: TCTA. Personal communication and AMD Treatment Plant Site Visit, 09 April 2018. Springs, South Africa.

Trans-Caledon Tunnel Authority (TCTA). 2019. TCTA website. Available at: https://www.tcta.co.za/. Accessed on: 14 August 2020.

Tempelhoff, E. 2019. Minister rejects plan for giant coal mine in Gauteng. *Fin24*, 23 October. Available at: https://m.engineeringnews.co.za/article/minister-rejects-plan-for-giant-coal-mine-in-gauteng-2019-10-24. Accessed on: 2 December 2020.

Turton, A. 2018. Mine voids. Personal communication, 18 October 2018.

Van der Merwe, C. 2018. Chairman: Blesbokspruit Trust. Personal interview, 21 February. Springs, South Africa.

Van Niekerk, J. 2018. Consultant: Karan Beef. Personal interview, 15 February. Roodepoort, South Africa.

Chapter 8
AMD Treatment Can Rid Its Bad Reputation

Abstract This chapter presents the main conclusions of this book. This book presents an example of a case in which it is much more difficult for decision-makers to manage perceptions and social constructions in the absence of an environmental impact assessment (EIA). The perceptions dealt with the following: How various stakeholders and key individuals perceived the purpose of an EIA, considering the short-term treatment (STT) was implemented under emergency measures; whether the government had abided by the law in implementing the acid mine drainage (AMD) STT or whether authoritative power was used; and how the impact of new mining applications and on-going pollution incidents would be managed, given the already sensitive Blesbokspruit system. Perceptions also included whether the management of the Blesbokspruit would be improved to account for all the activities taking place that depend on water uses. Lastly, whether the eventual long-term treatment (LTT) would also be implemented as an emergency, without considering the public's concerns.

Keywords Perceptions · Social constructions · Vested interests · Water quality · Power relations

The main research question that directed this book was *how the water quality of the Blesbokspruit is socially constructed, in the context of acid mine drainage (AMD) and its treatment.* Three objectives were identified to guide the book in addressing this question: The first was to explore the different perceptions of various stakeholders and key individuals on the water quality of the Blesbokspruit in the context of AMD and its treatment. The second objective was to identify and explain the contributing factors, processes and contexts that play a role in the varied social constructions of the water quality of the Blesbokspruit. The last objective was to explore how the social constructions of water quality are linked to power relations regarding AMD and its treatment. This chapter presents the main conclusions of the book.

To reach the conclusions of this book, the following served as guiding points to identify how the perceptions of the various stakeholders and key individuals on the short-term treatment (STT) of AMD developed. Since no environmental impact

S. Naidoo, *Social Constructions of Water Quality in South Africa*,
https://doi.org/10.1007/978-3-030-98237-9_8

assessment (EIA) was conducted for the construction of the AMD treatment plant and the discharge of neutralised water into the Blesbokspruit or for the disposal of sludge into the mine void, there was no common ground on how perceptions could develop. This book presents an example of a case in which it is much more difficult for decision-makers to manage perceptions and social constructions in the absence of an EIA. The perceptions dealt with the following: How various stakeholders and key individuals perceived the purpose of an EIA, considering the STT was implemented under emergency measures, whether the government had abided by the law in implementing the AMD STT or whether authoritative power was used and how the impact of new mining applications and ongoing pollution incidents would be managed, given the already sensitive Blesbokspruit system. Perceptions also included whether the management of the Blesbokspruit would be improved to account for all the activities taking place that depend on water uses. Lastly, whether the eventual long-term treatment (LTT) would also be implemented as an emergency, without considering the public's concerns.

8.1 Perceptions of the Water Quality and the STT of AMD

Water quality was one of the key concepts in this book. Its social nature, as defined by this research, is the lens through which the perceptions and constructions were understood. The book has demonstrated how the social meaning of water and especially water quality is the product of the very nature of water. Water is one of the most basic needs of human beings. This statement on its own explains the social nature of water. Water is one of the elements of nature that is most prone to contamination or pollution because of the various water discharges for domestic and business uses but is also equally manipulated as it is treated or purified. The quality of water is therefore a key point in how humans relate to, and form their perceptions of, water.

In exploring the different perceptions of various stakeholders and key individuals on the water quality of the Blesbokspruit in the context of AMD and its treatment, an extensive review of the EIA documents for the Grootvlei Site 6/L/16 tailings storage facility (TSF) was conducted. This formed a crucial part of the research in order to understand various stakeholders' perceptions of the treatment of AMD. It is not that stakeholders regard the water quality of the Blesbokspruit as being poor, but they are knowledgeable about, and understand that, the many activities and uses of the water of the Blesbokspruit contributed to a deteriorated system. This background knowledge and understanding influences their definition of water quality, in conjunction with their stakeholder group. This means that their need for, and use of, water, or their institutional role, influences how they define water quality. This all means that their concerns about water quality already existed before the building of the AMD treatment plant. This book has demonstrated that these perceptions are based on the potential impacts that the STT would have on the quality of the water in an already exacerbated system. There are a variety of stakeholders who interact with the Blesbokspruit and their perceptions of the water quality differ.

A common perception was that the effects of untreated AMD flowing into the Blesbokspruit would have been far worse than the current partially treated AMD. However, still, negative perceptions of the STT of AMD developed. A lack of trust already existed between the public and the Department of Water and Sanitation (DWS) because of how the EIA process was handled. During the EIA public participation process, stakeholders suggested that the mine void be used but it was regarded as being flawed due to possible impacts on groundwater. The mine void then became the sludge disposal site, and because it was implemented under emergency measures, the DWS was exempted from conducting an EIA. Only once the treatment was implemented did the public become aware of the new site. This enabled them to raise concerns about its impact on groundwater and what the purpose of an EIA was. To various stakeholders, this meant that the public participation process had no merit, because their views and suggestions had not been taken into account. The transfer of information and lack of communication to the public regarding the STT led them to question the management and legitimacy of the STT, and the decisions made were linked to a power play. Their use of the water and their dependence on it are influenced by their particular interests, which influence how they perceive the water quality, due to the implementation of the AMD STT. Such a case has the potential to alter how people define water quality.

8.2 Social Constructions of the Water Quality

The contributing factors, processes and contexts that played a role in the varied social constructions of the water quality of the Blesbokspruit were identified and emerged through the data analysis stage. The book concludes that AMD STT had developed a "bad reputation" based on its perceived impacts on the water quality of the Blesbokspruit. This was due to factors such as vested interests that influence social constructions and how these interests are actually embedded in power relations. An important conclusion in this book is that water quality is perceived and influenced by several factors. The constructions are therefore not the outcome of an objective assessment of the water quality but rather a scientific assessment of it. This implies that the different individuals understand what type of water is acceptable for them based on their water needs and interests. This book has shown that water quality is therefore not understood only through objective or scientific measures but is a product of perceptions in the sense that the uses of water are more important than its objective qualities. The choice of social constructionism as the framing for this book was influenced by the social science premise that human-created events in the Blesbokspruit area led to socially produced understandings of water quality in a particular way. This case study has demonstrated how the stakeholders' and key individuals' understanding of water is rooted within social relations and realities.

The constructs of the key individuals interviewed for this book came from influences that impact on their thinking about water and its quality. More importantly, the book concludes that these constructs are entrenched in power relations. The following conclusions about social constructions of the water quality of the Blesbokspruit and their link to power relations are presented in response to the main research question that guided this book.

8.2.1 *The Physicality of the Blesbokspruit*

The book has concluded that the water of the Blesbokspruit was physically reconstructed over the years to serve dominant social interests, such as the mining industry and, essentially, that of the government for economic growth in South Africa. It is widely known among various stakeholders that the water of the Blesbokspruit is physically co-produced by different sources: the rainfall, the underground water accumulating as a result of the mining industry and the water from wastewater treatment works and other sources.

The Blesbokspruit has been flowing constantly due to the years of discharge of mine water and wastewater treatment works. This flow is now enhanced by the large discharges of neutralised water from the STT of AMD. The increased volumes and flow of water are therefore a result of human interventions. These human-induced changes that occurred over the years meant that the physicality of the river has changed from a seasonal flow to a constant flow. This illustrated how the water of the Blesbokspruit can be seen as social. The contribution of these human interventions to the change in the physicality of the Blesbokspruit supports the relevance of social constructionism as the framing for this research within the specific context of AMD and its STT. Therefore, this Blesbokspruit case study confirms the link between the social and natural sciences, between human beings and nature and between objective conditions and perceptions, which are also key elements of social constructionism.

The book concludes that, depending on different interests, there are differences of opinion about the changes in the flows and volumes of water in the Blesbokspruit's hydrological physicality. The differences of opinion are about how these changes affect their interests, and therefore, their opinions and perceptions stem from the importance of these changes. Key individuals from the community and tourism sector found the higher volumes of water from the STT discharges to be beneficial for the system. This was due to the increase in bird species and because it provides a dilution effect for sewage, the smell from sewage spills impacted negatively on tourism. However, the way in which the reed encroachment was managed by the different government departments problematised these positive views, such as that it impacted negatively on the interests of the tourism sector, the residential community along the river and the environmental (birders) community. Therefore, constructions about flows and volumes of water due to changes in the physicality of the river were based on the management of sewage and reeds. This construction of the water quality was linked to power relations that stem from lack of resources or capacity to manage the water effectively. Government departments exerted their authoritative power by only managing areas that would reduce the flooding and only allowed permission to the community to assist after complaints had been raised.

8.2.2 Vested Interests

This book concludes that vested interests are a particularly significant factor in how water quality is socially constructed. Those who have lived in the Blesbokspruit area for decades were influenced by their vested interests in order to understand the consequences of pollution from mining and therefore reached the perceptions that they do. For instance, for those with an environmental interest in the area, the Marievale Bird Sanctuary was an important factor in understanding how their interests would influence their constructions. The sanctuary has thrived over the years and become a popular tourist attraction. However, the prominent years of gold mining negatively impacted on the bird species. Now that the water is flowing constantly and is of better quality, it has a beneficial impact on the birds. In this instance, the way individuals viewed the water quality depended on its impact on the bird life. The impact on the bird life influenced constructions for property owners along the river, which were more negative because it led to reed encroachment and occasional flooding of their properties. Their interest in their properties maintaining a profitable value was negatively affected by the higher volumes of water. This leads to the conclusion that constructions and perceptions might not always be logical in reality but are based on social understandings that are influenced by interests, which change on the basis of needs.

8.2.3 Mining

The historical context of mining in the Blesbokspruit area influenced key individuals' constructions based on their knowledge of mining and how it had impacted on the water quality of the area. Key individuals have either worked or lived on the East Rand for many years, which enabled them to understand the dynamics of the area. Individuals residing on the East Rand have seen how mining thrived and contributed to the economic growth of South Africa, but they have also witnessed how the decline in mining resulted in abandonment of mines, which led to the excessive environmental degradation. They have experienced how the government did not hold those mines accountable for environmental degradation.

Key individuals' understanding of the historical context of mining on the East Rand led to perceptions of the impact of current and future coal mining. Though AMD in the Witwatersrand is primarily associated with gold mining, the recent development of new coal mines in the Blesbokspruit area becomes cognitively linked to the historical legacy of mining in general in this area. Coal mining therefore plays a strong role in influencing social constructions of the water quality. Constructions in the case of the Blesbokspruit were based on the impact of coal mining on water quality and its encroachment on agricultural land that could be used for food production. Depending on the use of, and interest in, the water, constructions of the water quality differed. Agriculture in comparison to coal mining was referred to as the "lesser of two evils", even though mining created excessive wealth, economic growth and employment in South Africa. The way coal mining influenced constructions depended on who viewed it as the "lesser of two evils" and for what purpose. Those who favoured agriculture, other than the sector itself, were those who resided in the area and those in the tourism sector, because coal mining has the potential to further degrade the environment. This comes from the knowledge of the government not holding mines accountable in the past. However, those individuals who lived in townships seemingly favoured the mining initiative due to its potential employment opportunities. Further, key individuals such as those from the DWS and the Trans-Caledon Tunnel Authority (TCTA) who are involved in managing the AMD STT also opposed coal mining, in the interest of preventing further pollution of the water, which they are currently treating. The government is responsible for regulating both sectors. The social constructions of water quality are also different among government entities such as the DWS, the Department of Mineral Resources (DMR), the Department of Environmental Affairs (DEA), the Gauteng Department of Agriculture and Rural Development (GDARD) and the City of Ekurhuleni (CoE). These differences were based on the institutional interests of the respective departments.

Therefore, negative perceptions developed based on how the impact of new coal mining applications and ongoing pollution incidents would be managed, given the already sensitive Blesbokspruit system. The way coal mining influenced individuals' social constructions were linked to power relations because there were issues between spheres of government – national government departments making decisions through authoritative and legal power, without communicating and informing provincial and local government who are responsible for managing and monitoring the water, and then within the national government, and the decisions made to issue licences and authorisations based on their portfolio, and therefore what is in their institutional interest. In retrospect, lack of communication between and within spheres of government was problematic and impacted the management of the Blesbokspruit. How the Blesbokspruit was managed influenced how individuals constructed the water quality based on power realities that stemmed from this lack of communication and information.

Concerns regarding how the Blesbokspruit was managed often surfaced and questions about whether the management would be improved to account for all the activities taking place that depended on water uses were raised. Those taking decisions must understand that stakeholders have vested interests, and some of these interests differ among stakeholders and among government departments, such as the DMR, the DEA and the DWS, because of their respective institutional interests and obligations. The challenge lies in how the government manages these interests.

It was not possible to include all interests when implementing the STT of AMD, but some of these interests could have been beneficial to the management of the Blesbokspruit, because they come from people who care about the environment and the quality of the water of the Blesbokspruit. People who have lived in the area for years know and understand their surroundings; their interests are also beneficial to the environment and, consequently, they become experts in the process. The only way to include all these interests is to communicate and share information widely, especially to those whose interests may be impacted on.

8.2.4 Technology, Agriculture and Water Quality Sampling

Within the agricultural sector, there are differences of opinion on the impact of neutralised (AMD) water on agricultural activities. These differences are the product of different vested interests and water uses in the agricultural sector. As a result, these interests are responsible for how the constructions of the water quality are made. What was not commonly known to stakeholders outside the agricultural sector is that neutralised water is good for certain crops that require high levels of nutrients. Cattle need large quantities of water daily, but water that contains higher sulphate levels will have a negative impact on them. High saline levels in water impact negatively on the production of fresh vegetables and therefore subsistence farmers have a negative construction of that kind of water. However, certain crops such as beans need a high nutrient intake and therefore views on neutralised water differed. Significant as a conclusion is how this example of the agricultural sector demonstrates that differences in interests even within a sector can produce differences in social constructions about water quality.

Another relevant factor is water quality sampling. The book identified clear differences in opinion about the accuracy of sampling. Concerns were raised about the accuracy and consistency in sampling. What is meant to be a scientifically credible process was not credible for some due to the way samples were taken. This influenced how constructions of water quality were formed.

Though water pollution was not directly linked to AMD treatment, in practical terms it is very difficult to separate the impact of the treated water from the effect of pollution on the Blesbokspruit. Though they should not be conflated, in practice, the research found that it did happen. Lack of official enforcement of anti-pollution measures created the perception with some individuals that the government lacked the concern to address the ongoing pollution. This book can conclude that whether action was taken depended on how influential the individual reporting the incident was. It also depended on the capacity and resources of the government to enforce compliance. It can be concluded that anti-pollution measures provide an indication of the role of power and authority in social constructions of the water quality, even if they do not deal directly with AMD and its STT.

The use of technology and processes followed in the Blesbokspruit have played a definite role in how key individuals constructed the water quality. This refers to pollution-related activities (e.g. sewage mismanagement), the effect of the herbicides used for reed spraying on the water quality, the process of water quality sampling and the perceived impact of neutralised water with high sulphate levels on agriculture. Therefore, the combination of reed spraying, sewage spills and the STT discharges could exacerbate the negative impact on the water quality and does influence social constructions.

8.2.5 Information and Communication

Information and communication strongly influenced the social constructions of the Blesbokspruit water quality. More so, it was strongly embedded in power. The findings of the research indicate that the credibility of the AMD treatment project depended on the sharing of information with the different stakeholders, as well as taking their views into account in the planning phase. The fact that the approved EIA for the sludge disposal site became obsolete because another site was used without an approved EIA created a serious trust deficit with the public. An EIA public participation process serves as an important medium for the public to be heard and for communication and information sharing. This became evident in the responses from stakeholders when they were not informed of the fact that the mine void, and not the approved site, would be used for sludge disposal when the STT was implemented. Though it may not be factually correct, the social constructions that emerged from this situation suggest that people thought that they had been deceived by the EIA process and deliberately kept in the dark afterwards. This meant that stakeholders who were part of the STT planning phase felt that they had not been heard. The negative perceptions about the STT that developed during the EIA phase led to negative constructions of the water quality, which were influenced by the use of communication and information. This book concludes that the STT being implemented under emergency circumstances means that the relevant information was not shared widely and the decisions made were viewed by stakeholders as serving certain interests, such as the AMD project team, and not the community. The book concludes that differences in access to information played a major part in why perceptions of the STT differed so much. Access to information also coincided with different vested interests and differences in the power relations between the stakeholders. Those who had knowledge of the implemented STT process felt it was beneficial for the system, but this sentiment was not shared among the other stakeholders. Those with access to the information became the "insiders" and those excluded from it, the "outsiders".

The relative nature of information and its impact on social constructions was also revealed in this book. During the EIA process of investigating the impact of the sludge disposal above ground, several of the stakeholders suggested the use of the mine void as their preferred site, but this was rejected. However, when surprisingly the mine void was used in the end, albeit without an approved EIA, most of the so-called outsiders' social constructions were very negative. The conclusion therefore is that information on its own (or the nature of the information) is not the decisive factor but how that information is communicated, how accessible it is and how the information is used to justify a decision. It is well known that information can be manipulated. The credibility of information depends on how it is used and therefore how it can support particular social constructions.

Knowledgeable members of the community are an information-sharing grouping and they are often the only means of information for the rest of the community. The public believe that it is the government's responsibility to relay such information and keep them informed, because it is their constitutional right to be informed. In the government's absence, from the research it was evident that influential members of the public became credible sources of information and, as a result, developed a form of power. Key individuals from the DWS suggested that communication needed to be more transparent through the Blesbokspruit Forum, but when additional questions were asked about the STT, the LTT or water quality sampling, at times information remained confidential. This suggests that authority and a power play between government departments took precedence in the sense that some were seen as the real decision-makers, while others were merely reporting the decisions to the public.

This book has demonstrated the importance of the Blesbokspruit Forum as an instrument of public communication and information sharing. For a government project such as the AMD STT, it serves as an example of how public-private interests can merge and how public participation can be structured, and its role as an influencer of social constructions is undeniable. However, the forum could have been used as a means to strengthen communication and information sharing, which was still questionable at the end of the research. An official in a national government position can use the forum to manage perceptions and constructions of the water quality by effectively communicating and sharing information. It could also take advantage of the fact that many representatives in the forum are from civil society and the private sector and are willing to assist with their expert knowledge. Community members also attend the forum to report incidents. Therefore, they are a crucial source of information for the government to better its task of managing pollution and illegal activities.

The overarching conclusion about the factors that influenced the constructions of water quality is that the facts about the water quality in the Blesbokspruit do not speak for themselves. What counts as truth about water quality varied, depending on the perspective of who was talking about it, his/her purpose and what his/her individual interests were. It also depends on whose decision is the most influential, whose decisions are based on authority or legalities and whether these decisions are impeded by lack of resources or by capacity issues. At the same time, decisions are linked to power relations between individuals, institutions and society.

8.3 Water Quality and Power Relations

Using social constructionism as the framing for this book assisted with explaining how individuals understand water and its quality. Power is an important element of these relationships. Power suggests the presence of hierarchies and inequalities, privilege and marginalisation, influence and acceptance. The impact of a person or institution on the formation of social constructions is in part influenced by their position or status of power. Those stakeholders with more power means more of an influence on their social constructions. Particular interests within the water sector make one think differently about power relations between those who manage the resource and those who use the resource. The interviews and other sources used for this research serve as a valuable tool on how power can be analysed in social contexts in terms of the following elements: power as influence, power as authority or legal power, power determined by communication and information, capacity and power as a product of credibility and reputation.

8.4 AMD Treatment Can Improve Its Reputation

This book has shown how people's understanding of nature is rooted within social relations by explaining how the Blesbokspruit was physically reconstructed over the years to serve dominant interests, for instance, the mining industry and that of the government for the economic growth of South Africa. The book contributes to an explanation of how the meaning of water emerges from relations with, and through, water. How the water quality of the Blesbokspruit is socially constructed indicates how one relates to water, based on one's need and use for it. For example, interests of key individuals changed from that of mining due to being employed by the industry to environmental interests and increasing the bird life by preventing new mining from taking place. The treatment of AMD has not been researched from a social science perspective. This is informed by the perceptions that stakeholders have on the STT of AMD due to the unknown possible social and environmental impacts that it could pose to the water quality. The book has illustrated how the constructions of water quality were formed on the basis of how the treatment of AMD was perceived.

This book contributes towards understanding how relationships through water are grasped through power relations (regarding AMD and its treatment) and vice versa. How water quality is constructed is linked to power relations between those who manage the resource and those who use the resource. In relating ideas about water quality to power dynamics, this book contributes to a critical understanding of relations between society and nature, especially water and its quality. This book is unique compared with other studies conducted on water and power relations in South Africa because it explains how the treatment of AMD was exempted from legal requirements in order to protect the water quality of the Blesbokspruit and also how power was used to influence decisions to treat the water to prevent further deterioration. This led to AMD treatment gaining a somewhat bad reputation, because this included restricting information and making decisions based on legal and authoritative power. However, the interests of individuals led to mixed perceptions of how power can influence the treatment of water. Legislatively allocated power determined the role that the different spheres of government played in the AMD STT and how it influenced how water quality was socially constructed by individuals from civil society, business and the community. Formal processes, such as the EIA and those followed in the treatment of the mine water, were also perceived as the product of power relationships, and stakeholders' assessment of these relationships contributed to whether their constructions were positive or negative.

The book identified elements of power that exist in how water quality is managed in the context of the AMD STT. Power in the form of authoritative or legal power in relation to reputational power plays a part in specific aspects of the construction formation processes. A case in point was the question whether an EIA and water use licence for the AMD STT in the Blesbokspruit case study were required by law or whether a declaration that an emergency situation existed could be regarded as a legitimate exemption. Further, if a project is granted an emergency status which exempts the project from conducting an EIA (as in the case of the AMD STT project), it should not mean that all initial stipulations in legislation (such as National Water Act (NWA), The Constitution and the National Environmental Management Act (NEMA)) are no longer adhered to, such as people's constitutional right of access to information. Some individuals are extremely knowledgeable about legislation, and even though a project can be exempted from certain legislative requirements, due to valid reasons, to them some of the original stipulations in these Acts should still be adhered to. This is because those stipulations do not affect the implementation of the project from a natural science perspective (the actual treatment) but from a social science perspective (the social impacts). Social constructionism as the framing for this book was therefore significant because individuals felt that the government did not abide by the law when implementing the STT, which influenced their constructions of the water quality and contributed to their negative perceptions of the STT. The book showed that the different government departments regarded their legal power as sufficient to justify such an exemption. Other stakeholders questioned that assumption on the basis of their constructions that no such legal power existed. This illustrates the point that social constructions influenced by power considerations often have little to do with the objective characteristics of water quality but much more to do with circumstantial or contextual influences such as power relationships.

The conclusions of this book have important implications for institutions and processes of water management. The findings have a direct impact on how public opinion is formed, how the public responds to public policies on water management and how a sustainable development equilibrium can be established between water and different vested interests in society.

The following lessons can be learned and changes implemented: First, in order to strengthen the aims of co-operative government, water use licences (DWS), environmental authorisations (DEA) and mining rights (DMR) for each applicant/user should be conducted as a concurrent process to determine the cumulative impact on the water course. This will ensure that all government departments responsible are aware of the licences and applications that are issued. The main aim is to ensure that no further harm or threats are posed to the Blesbokspruit, especially since extensive rehabilitation in the form of the STT of AMD is taking place.

Second, despite the fact that the STT of AMD can be regarded as a much better option than no treatment at all, it did not account for all the users or managers of the water. This therefore has policy implications for the DWS. When implementing the LTT there is the potential to rectify the implications that stem from power relations present during the STT. The first step is through strengthening communication and information sharing. The next is that the EIA public participation processes should be taken more seriously in the future. This means that the process should include all stakeholders involved by listening to their views, ensuring that there are expert responses to these views to avoid incorrectly informing the public, and then adapting the relevant public suggestions on the process, which is part of the purpose of an EIA. Clarifying roles and responsibilities of government departments will strengthen this process and achieve the aim of co-operative governance. If all this is taken into account, perhaps more receptivity and support will be given to the government in its pursuit of better water management. This will also produce more positive constructions of the water quality of the Blesbokspruit.

Further, through this case study, this book aims to contribute to the global discussion on mining, especially regarding how lessons can be learned with regard to policies, practices, water management and public participation processes.

Index

S. Naidoo, *Social Constructions of Water Quality in South Africa*,
https://doi.org/10.1007/978-3-030-98237-9